SpringerBriefs in Electrical and Computer Engineering

More information about this series at http://www.springer.com/series/10059

Christian K. Karl • William Ibbs

Developing Modular-Oriented Simulation Models Using System Dynamics Libraries

 Springer

Christian K. Karl
Department of Civil Engineering
University of Duisburg-Essen
Construction Technology Program
Essen, Germany

William Ibbs
Department of Civil and
    Environmental Engineering
University of California
Berkeley, CA, USA

Ibbs Consulting
Oakland, CA, USA

ISSN 2191-8112   ISSN 2191-8120   (electronic)
SpringerBriefs in Electrical and Computer Engineering
ISBN 978-3-319-33167-6   ISBN 978-3-319-33169-0   (eBook)
DOI 10.1007/978-3-319-33169-0

Library of Congress Control Number: 2016938639

Printed on acid-free paper

This Springer imprint is published by Springer Nature
The registered company is Springer International Publishing AG Switzerland

Preface

The development and application of simulations with the purpose to explore systems and their innate, partly dynamic structures has established itself, among others, in the areas of natural sciences, business studies and political sciences. This similarly applies in the field of education to support the understanding of complex relationships, especially in business studies. Both always have a model in common which serves as starting point for the development.

The aim of this book is to introduce the development and practical application of a module-oriented development framework for domain-specific system-dynamic libraries (SDL approach), which can be used in the simulation of multicausal and dynamic relationships on different levels of an industry, as an example the construction industry. The book will enable both academics and practitioners to develop the first systems right from the beginning. The successful conceptual design of this development framework demonstrates that it is quite reasonable and possible to connect the development of simulation models and daily work.

Thereby, knowledge synergies will be created which enable the interdisciplinary development of simulations in the sense of a synergistic knowledge absorption (SKA method). Multidisciplinary research and development teams and scientists from different domains, as well as practitioners, have numerous possibilities to develop SDL units from varying perspectives based on this approach. Compared to other fields, significantly less implementations of simulations exist in the construction industry. Therefore, the approach introduced here provides a valuable contribution to promote further developments, e.g. the explanation of the risk situation of a company, the identification and evaluation of project risks and endangered operational procedures on various functional levels, or to improve the understanding of the decision-making process in detail. Nevertheless, the introduced approach is suitable for any kind of business, independent of decision level and functional area.

The authors would like to acknowledge the support of the following persons: Christian K. Karl owes deepest gratitude to his whole family, especially his beloved wife Maja and their children Sophia and René. Without the essential support in the background, he would not be able to do research in such manner like presented here.

William Ibbs appreciates the support of his wife, Kerry, and their children, Charlie, Courtney and Zach, his parents, his students and colleagues and his consulting clients. They've all made him a better person and their influence is sprinkled throughout this book.

Essen, Germany   Christian K. Karl
Berkeley, CA, USA   William Ibbs

Contents

List of Figures

List of Tables

Nomenclature

AMCA	Atom-molecule-component approach
BIM	Building information modelling
BRTV	Federal Collective Agreement for the Building Industry (Bundesrahmentarifvertrag für das Baugewerbe)
CCD	Construction company dynamics
CD	Construction dynamics
CDL	Construction dynamics library
CLD	Causal-loop diagram
CMD	Construction market dynamics
CPD	Construction project dynamics
CPFS	Construction project flight simulator
ERM	Entity relationship model
GUI	Graphical user interface
LC	Lean construction
MIT	Massachusetts Institute of Technology
MLA	Multilevel analysis
PGM	Probabilistic graphic models
PLE	Personal learning edition
PO	Process Ontology
SD	System dynamics
SDL	System dynamics library
SFD	Stock-low diagram
SKA	Synergistic knowledge absorption
SMP	Shared modelling platform
UML	Unified modelling language
USDML	Unified system dynamics modelling language
VDI	Association of German Engineers (Verein Deutscher Ingenieure e.V.)

Chapter 1
Introduction

This section explains the fundamentals which are seen as indispensible for the understanding of the content of the book. However, this chapter does not claim to offer a comprehensive and highly detailed overview of the issues, but aims at introducing the relationships of the different domains to make them available for the further chapters of this book. In the following the motivation and the general aim of this book are described, followed by central definitions and characterizing features. If the reader is interested in gaining more details about specific topics, the authors suggest to have a closer look into the cited literature and the further readings, which are listed at the end of the book. The authors also welcome personal feedback and dialogue.

1.1 Motivation and Aim

Simulation models are complex descriptions of reality to aid the exploration of multicausal processes within a system. The development of a simulation model, necessarily very detailed due to the included structural relationships and interdependencies, has to consider numerous aspects. As both human resources and financial expenses required for the development and implementation of simulations can be quite substantial, it appears sensible to support one of the most complex phases—the modeling.

Especially in this context, the development framework for system-dynamic libraries (SDL) offers the following advantages:

- A specific problem-oriented simulation model can be easily designed through combining already existing entities out of the SDL.

© The Author(s) 2016
C.K. Karl, W. Ibbs, *Developing Modular-Oriented Simulation Models Using System Dynamics Libraries*, SpringerBriefs in Electrical and Computer Engineering,
DOI 10.1007/978-3-319-33169-0_1

- Especially new, interested but so far unexperienced simulation designers will gain a direct access to the modeling process without the necessity to deal in detail with the theory first.
- A module-oriented model allows the design of variable and versatile simulations. Hence, the developed simulations are reusable for other problems by adjusting, adding or reducing the model.
- Such simulations can be executed on the basis of the complete modeled reality depicted therein or part of it.
- Due to the reason, that different simulation models will have a common basis, insights gained from one simulation scenario can be applied to another simulation scenario and vice versa.

Consequently, the development of an SDL offers considerable and numerous possibilities of application. This, in turn, results in the following central aim of this book:

The aim of this book is to equip the reader with a method with which he will be able to develop module-oriented simulation models, which can be used in the simulation of multicausal and dynamic relationships on different industry levels as well as in research, academic education and further training and practice. The intended audience is both the academic and the practitioner.

1.2   Definition of the Term "Simulation"

Generally stated, a simulation allows to emulate selected events or whole systems in a simplified manner (Wenzel 2004). Here the primary aim is more to answer questions than to understand processes (Taylor and Walford 1974). Due to the potentially high degree of complexity, a simulation is devised to illustrate systems, the relationships and the multiple interdependencies between the included individual elements in detail. This maximum of realism is generally independent of the individuals who use the simulation (Kriz 2000).

These relationships are also, among other sources, found in the VDI (The Association of German Engineers) guideline "Simulation of logistics, material flow and production systems" (in German: Simulation von Logistik-, Materialfluss- und Produktionssystemen), in which simulations are understood as the emulation of a system with its dynamic processes in an experimental model to achieve insights which can be transferred to reality (VDI 3633-1 2010).

In summary, the term simulation describes the precise emulation of a real situation with the aim to gain insights, but without the intention to serve as teaching method directly (Regardless of the fact that it is under certain circumstances also applicable for this purpose.).

1.3 Characterizing Features

Based on the aforementioned definition, the fundamental terms like model, system and simulation as well as their sub terms are to be considered further and need to be put in a context and relation to each other. As the concept of a model is the central item, it is necessary to discuss it in more detail in the following.

1.3.1 Model

The term model is ambiguous (Giesen and Schmid 1976) and is used frequently as well as with multiple meanings (Harbordt 1974). In general, every emulation of an original can be labeled a model (Dörner 1984). Models serve the purpose of cognitive insight (pragmatic function) on the one hand and inform about certain relations regarding an existing or future original (semantic function) on the other hand (Busse 1998).

Even though numerous definitions are potentially available, e.g. due to the fact that the term model is exposed to different viewpoints in the domains of engineering, economics, psychology, philosophy and other sciences, the definitions offered by Stoff (1969) and Stachowiak (1973) are seen as the most commonly accepted ones because of their relatively general character.

According to Stoff, models can be categorized into material models and cognitive models. Stoff understands material models to be e.g. analogy models, structural models, models with mechanical, dynamic, kinematic and other physically similar representations of an original as well as drafts and diagrams which show a spatial similarity to the original. Cognitive models, however, are drawing models (symbolic models), graphic presentations, maps, sketches and also visual (iconic) models, according to Stoff.

Stachowiak classifies the term model into graphic and technical models and semantic models. In graphic models, which are predominantly defined by two-dimensional descriptive and spatial images of the original (Stachowiak 1973), photographs, illustrations, diagrams and different types of graphs depict relationships and interdependencies within an overall structure. Currently, such models are applied as probabilistic graphic models, in which every knot represents one random variable (or a group of random variables) and the connections between these variables express probabilistic relationships (Bishop 2006).

Although this categorization appears similar to the material models following Stoff, it differs in the aspect that Stoff explicitly classifies images as cognitive models if they are not strictly true to the original whereas Stachowiak regards these as diagrams, figures or graphs in the widest sense.

The mainly three-dimensional technical models, according to Stachowiak, are to be seen as spatiotemporal and material-energetic representations of originals (Stachowiak 1973). One potential use of these models is the model of a flying body for studies in a wind tunnel, testing facilities for the testing of materials like the life span and durability of road surfaces made of different materials (Stoff and Stachowiak agree in this).

According to Stachowiak semantic models are abstract, formal descriptions and representations of an excerpt taken from reality. The transition of the graphical as well as the technical model to the explicitly semantic model is smooth and flowing (Stachowiak 1973). Hence, this model is found in, e.g., software engineering as a preliminary step before the concrete technical implementation in the data modeling. Different modeling languages exist for the creation of such semantic data models, of which the entity relationship model (ERM), which can be used as basis for a standardized viewing of data (Chen 1976), is the most widespread. Therefore the semantic models following Stachowiak can be seen as synonymous to the cognitive models proposed by Stoff.

Models are principally defined by three fundamental features: (1) the visual representation, (2) the reduction and (3) the pragmatic feature.

Regarding the visual representation, it can be stated that models are always representations of natural or artificial originals (Stachowiak 1973). Depending on the according purpose of the realization, models therefore are required to be sufficiently similar to the original system (Sauerbier 1999). Usually models do not comprise all properties of the original they represent (Stachowiak 1973), as stated by the feature of reduction. This means that less important attributes of the original are omitted by the model designer and the original is thereby narrowed down to typical and relevant influence factors, data, properties, events, information, structures and so on. The pragmatic feature expresses that models cannot always be allocated to an original as it exists in reality, because the model fulfills its substituting function with a limited scope on selected theoretical or real operations (Stachowiak 1973). Consequently, the model can contain elements which are not given in the original.

Salzmann expands these considerations. He added the features accentuation, transparency, intentionality and instrumentality to the model (Salzmann 1976). While the first two of the listed features contain only minor elements of innovation, the feature of intentionality suggests that a model is always developed for a specific aim. For Salzmann this aim can be teaching or education, but also a forecast of the development of future market conduct. The feature of instrumentality includes this consideration. It expresses the idea that each model follows an intention or serves a function. Buddensiek et al. (1980) focused especially on the intention of education and training and assigned the following training functions to the models:

- elucidating specific nontransparent structures of an original (structuring function),
- clarifying complex relationships (heuristic function),
- making learning content available which is not accessible in the original (substitute function),
- introducing alternatives to an existing reality (anticipation function),
- questioning an available reality from a real-utopian perspective (function of critique of ideology),

- offering the learner a training possibility for a specific behavior needed in future situations (training function)

Models are structured in themselves by allocating each subject a structured set of potential attributes, i.e. characterizing features. Attributes here mean characteristics and properties of individuals, relations between individuals, properties of properties, properties of relations etc. (Stachowiak 1973). Attributes are linguistically expressed with predicates (Stachowiak 1973).

Hence, abstract statements can be expressed as mathematical functions in models on the basis of attributes respective predicates. As a whole, these statements can form and display a system. Similar relationships are communicated by in the U.S. Army Modeling & Simulation Glossary: A physical, mathematical, or otherwise logical representation of a system, entity, phenomenon, or process. This background is applied in the conception of models, Chap. 2 starting p. 9.

1.3.2   System

Attributes and predicates represent so-called process variables in a system. According to Forrester (1972), a system is a multitude of parts (components) which are in relation to each other and interact for a common purpose. Kramer and Neculau (1998) postulate a similar view, stating that it is only possible to call something a system when several process variables are connected to each other and form an ordered whole. Usually it is possible to find one or more process variables which can be identified as independent values and others which are dependent values and influenced by the former (Kramer and Neculau 1998).

Systems are normally illustrated with the help of larger quantities of attribute and predicate classes (Stachowiak 1973). Thus, a model can also be seen as a system (Harbordt 1974). Each system is described with the help of the environment, the behavior function and the structure (Kaaz 1972).

The elements of a system can be seen as systems themselves, even on a lower level (Krüger 1975), i.e. they form a subsystem. Then a system can consist of several subsystems which in turn contain a certain number of elements respective units. If components cannot be divided any further, they are seen as elements (Niemeyer 1977).

The interaction within a system is effected by the links between the elements contained in the system, i.e. they are connected to each other by a characteristic system structure (Bossel 2004). The relationships between the individual components influence the system as a whole and, depending on the behavior pattern of the model, create variable impulses.

Summing up, systems are composed of the following parts: (a) elements, (b) subsystems, (c) relationships between the elements respective the subsystems.

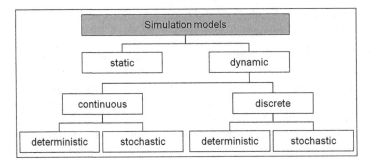

Fig. 1.1 Simulation models

1.3.3 Simulation

Following Sauerbier (1999), a simulation is the execution of calculations in a model, transforming input values into output values. Based on the previously introduced terms, the simulation thereby represents an imitation of real processes using mathematical models. According to Page (1999) simulations can be classified in the following types:

- examination method,
- medium of representation,
- intended purpose and
- transition of state.

It is especially the classification through the transition of state which is of central interest (Fig. 1.1).

A static model is a reproduction with no changes of state. This means that such a model represents a system which displays reality at a specific point of time only. In the end such a system contains only constant attributes, so-called state constants (Niemeyer 1973).

In contrast to that, a dynamic model comprises time-dependent changes of state, so called state variables (Niemeyer 1973). Although a distinction can be made between static and dynamic models, the definition of the term simulation, which is communicated through the Modeling & Simulation Glossary of the (U.S.) Army Modeling and Simulation Office, seems to be contrary in this context. This definition considers especially the aspect of time: "A simulation is a method for implementing a model over time.". Based on this, only dynamic models should be a basis for simulations. Static models are entirely absent here and are not considered in this definition. As state variables change with the progress of time, it is possible to distinguish between continuous or discrete changes of state. Within the continuous model, which can contain one or more differential equations to depict the relation between the progress of time and changes of the state variables (Liebl 1995; Sauerbier 1999), state variables change continuously. In a discrete model, however, state variables change at specific points of time, i.e. event-oriented.

In a deterministic simulation the reactions to specific input are explicitly set (Page 1999) and the input values thereby clearly determine the simulation results (Sauerbier 1999). Lastly, the stochastic simulation describes the input values with the help of probability distributions (Page 1999).

References

Bishop CM (2006) Pattern recognition and machine learning. Springer, Berlin

Bossel H (2004) Systeme, dynamik, simulation: modellbildung, analyse und simulation komplexer systeme, 1st edn. Books on Demand GmbH, Norderstedt

Buddensiek W et al. (1980) Grundprobleme des Modelldenkens im sozio-konomischen Lernbereich, pp 92–122. Modelle und Modelldenken im Unterricht. Klinkhardt, Bad Heilbrunn/Obb

Busse R (1998) Was kostet design? Verlag form, Frankfurt a.M.

Chen PPS (1976) The entity-relationship model - toward a unied view of data. ACM Trans Database Syst 1(1):936

Dörner D (1984) Modellbildung und simulation, pp 337–350. Sozialwissenschaftliche methoden. Oldenbourg, München

Forrester JW (1972) Grundzüge einer systemtheorie. Gabler, Wiesbaden

Giesen B, Schmid M (1976) Basale soziologie: wissenschaftstheorie. Wilhelm Goldmann, München

Harbordt S (1974) Computersimulation in den sozialwissenschaften. Rowohlt, Reinbek (bei Hamburg)

Kaaz MA (1972) Zur formalisierung der begriffe: system, modell, Prozeß und Struktur. Angewandte Informatik (12):537–544

Kramer U, Neculau M (1998) Simulationstechnik. Hanser, München

Krüger S (1975) Simulation: grundlagen, technik, anwendungen. De Gruyter, Berlin

Kriz WC (2000) Lernziel systemkompetenz. Vandenhoeck und Ruprecht, Göttingen

Liebl F (1995) Simulation: problemorientierte Einführung, 2nd edn. Oldenbourg, München

Niemeyer G (1973) Systemsimulation. Akademische Verlagsgesellschaft, Frankfurt a.M.

Niemeyer G (1977) Kybernetische system- und modelltheorie: system dynamics. Vahlen, München

Page B (1999) Diskrete simulation. Springer, Berlin

Salzmann C (1976) Unterrichtsmodelle, pp 449–453. Handlexikon zur erziehungswissenschaft. Ehrenwirth, München.

Sauerbier T (1999) Theorie und praxis von simulationssystemen. Vieweg u. Sohn Verlaggesellschaft, Braunschweig

Stachowiak H (1973) Allgemeine modelltheorie. Springer, Wien

Stoff VA (1969) Modellierung und philosophie. Akademie Verlag, Berlin

Taylor J, Walford R (1974) Simulationsspiele im unterricht. O. Maier (EGS-Texte), Ravensburg

VDI 3633-1 (2010) Simulation von Logistik-, Materialuss- und produktionssystemen - Grundlagen (Entwurf). Beuth Verlag, Düsseldorf

Wenzel F (2004) Bertelsmann Wörterbuch der deutschen Sprache. Wissen Media, Gütersloh

Chapter 2
Module-Oriented Modeling Approach

The main aim of this chapter is to introduce an approach for modular simulation models. In doing so, the models should be designed in a modular way, to flexibly allow for further developments. Building on separate components, a library will be set up, allowing for new elements to be embedded in it or for the existing ones to be adjusted or extended respectively.

For the simulation this is meaningful in two ways: insights can be gained both from using a single element and from the sensible combination of several interacting elements. This fact should also support the design of different simulations, in which distinct components can be sensibly combined in modules, following the logic of gaining knowledge. The module-oriented approach developed here is based on the System Dynamics method which is introduced briefly in the following chapter.

2.1 System Dynamics

System Dynamics (SD) is a system oriented and computer-based problem-solving approach for explicit mathematical models (Milling 1996), addressing the analysis and design of decision-making rules (policies). Jay W. Forrester is considered to be the founder of the System Dynamics approach. Working at the Massachusetts Institute of Technology (MIT) at the end of 1950s, he linked the Feedback Control Theory (for details see Doyle et al. 1992) with the Computer Science and Business Management (Hafeez et al. 2004). Forrester's research findings in the field of theoretical dynamics and in the systemic behavior of industrial companies were also applied in other scientific fields and thus the generic term System Dynamics was coined. This is the reason why it was published for the first time with the term 'industrial dynamics' (Forrester 1961) This approach applies to dynamic problems characterized by change over time, interdependence of system

© The Author(s) 2016
C.K. Karl, W. Ibbs, *Developing Modular-Oriented Simulation Models Using System Dynamics Libraries*, SpringerBriefs in Electrical and Computer Engineering, DOI 10.1007/978-3-319-33169-0_2

components, information feedback and cyclical causality. As these occur in every complex system, be it social, industrial, economic or ecological (Richardson 1991), they also apply to teaching, research and decision-making theory. The next sub-chapter offers a short overview of the background and the basics of this approach. More detailed descriptions and discussions of System Dynamics can be found in Forrester (1961, 1968, 1969, 1971b, 1972a,b), Sterman (2000) and Sterman (2004).

2.1.1   Background

SD is an approach that attempts to comprehend the behavior of complex systems over time. It addresses feedback loops and time lags affecting the whole system. For this purpose, quantitative and qualitative models are developed, which can, due to their inner structural properties and data, reproduce the real system under examination as accurately as possible.

As opposed to considering separate isolated variables and their development, the SD approach is focusing on the interdependency of various factors and variables over a time sequence. Compared to other methods, e.g. the econometric approach in which variables are given over the whole of a prediction period, an initial state is defined in SD by determining the starting points for all variables (Schwarz and Ewaldt 2002). This will dynamically change across time through manipulation of main parameters in form of endogenous pulses. Econometric models are part of empirical economic research and attempt to elucidate economic and structural relationships as well as test economic hypotheses (for more details refer to Eckey et al. 2011).

This is therefore a cyclical self-updating model throughout the simulation process (Weber and Schwarz 2007). Due to the causal links between the several variables, both dynamic as well as time-lagged interdependencies can be observed, thus allowing a deeper understanding of the system. In this respect, SD is a suitable method for the detailed analysis of complex dynamic problems. The analysis of the problem structure and of the reactions it causes provides for insights on long-term effective decision-making rules.

Because of its universal formalism, SD offers a wide range of applications in various fields. Especially the complex economic field requires such a systemic approach for an adequate modeling (Gold 2005). The SD method is therefore highly suitable for the modeling and simulation of markets, productions and companies.

The SD models clearly distinguish between stock- and flow-variables. Stock variables represent the current state of the system, for example the amount of produced goods.

In contrast, flow variables have a changing effect, for example the amount of produced goods per hour. All SD models consist mainly of these two sorts of variables, whereby all elements are inserted in the model with given values (initial values), so that the future system states can be processed.

Such a model is subsequently further refined, until the real system is captured as accurately as possible respectively needed. The computer-simulation of the system addresses the inner structure with regard to nonlinearities, feedback structures and complexity. The simulation of various interventions should deepen the understanding of the system under examination and conclusively explain decision-making rules for the system governance or for interventions. Nevertheless, this is legitimately only a model, meaning an abstraction of the reality. That is why, depending on the purpose of the simulation, the model should be designed so as to explicitly incorporate relevant aspects, whereas less relevant ones are to be simplified if not even disregarded. The expert knowledge of the modeler is therefore paramount for the design of such models (for further details regarding the simulation of systems, refer to Sterman 2004). Despite the necessary abstraction, the model should remain realistic and depending on its aim, should provide insights about the system structure even to SD-laymen. Consequently, the representations of the model should be formally clearly outlined, to facilitate the design, analysis and presentation of the real system part.

2.1.2   Excursus: Mental and Formal Models

In the process of modeling, a basic distinction between mental and formal models must be made (Fig. 2.1).

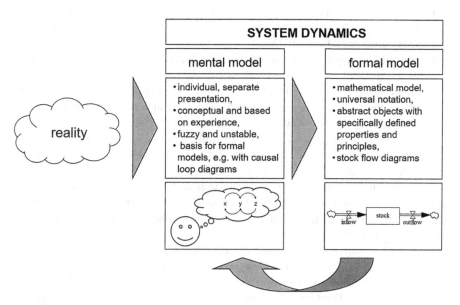

Fig. 2.1 Relationship between mental and formal models

The term mental models is often used to describe the cognitive structures of individuals and was coined in the early 70s by Johnson-Laird. He defines structures as inner, differentiated representations complementing the logic application and consequently the ability to draw conclusions (Johnson-Laird 1983; Baron 2000). Mental models are therefore to be considered as more or less exact thought patterns about the connection between specific situations (Johnson-Laird 1983). Senge (2003) understands mental models as inner representations of the nature of things, e.g. pictures, assumptions and stories that trigger one's own interpretation of the world and determine one's behavior. This is one of the reasons why Forrester criticized mental models as fuzzy, incomplete and imprecise (Forrester 1971a). Even more so, mental models change over time within an individual (Forrester 1971a), thus leaving out feedback, time lags and accumulations (Sterman 1994).

Doyle and Ford (1999) also have a similar perspective. According to them, the mental model of a dynamic system is a relatively durable and accessible but limited inner conceptual representation of an outer past, present or future system, the structure of which parallels the perceived system structure.

Even though, according to Strohhecker, mental models are constantly under plausibility controls by the learning processes, the latter render them rather unstable and are therefore to be seen as inaccurate and not (without information loss) interchangeable. Only to a limited extent can they be subject to examination and evaluation, as they offer only a rough estimate of the relationship between the influencing variables (Strohhecker 2008).

To sum up, mental models appear to be better suited for the representation of smaller systems. As they are largely based on an inner, individual and consequently mainly subjective perspective, an unbiased transfer and communication is to be regarded as problematic. This is not least because of the fact that mental models exclusively emerge and exist in the mind and therefore are frequently not formally documented. Consequently, the mental models themselves as well as the gained insights are difficult to reproduce. Especially when taking into consideration the state of the art quality requirements for scientific knowledge gain (among others objectivity, comprehensibility, verifiability, reliability and validity), the exclusive use of mental models is ill-suited for complex dynamic systems.

On the other hand, mental models are quite suitable to serve as a basis for the design of formal models (see Sect. 2.1.4.1). Formal models are primarily based on the mental representation of at least one individual, but are designed and documented following a formal approach and universal notations. Such a model consists of abstract objects with pre-defined properties, for which certain rules (axioms) apply (This aspect was already discussed in general according to Stachowiak (1973) on pp. 5).

The basic structure of a formal model can, for example, emerge from the predicate logic. Such a formally designed model results from reality either by abstraction or by detail accumulation and eventually becomes accessible for mathematical processing. Therefore, these models are more than an individual conceptual representation of reality.

Through the use of formalism, everyone can reach an understanding of the system structure, predict possible development trends, identify potential interferences and thus perform effective interventions in the system. Following this line of argument, formal models are to be preferred when examining complex systems, as they can assist the optimization of mental models and consequently the assessment and decision-making capabilities.

When modeling, one should follow Valéry (1937), to whom simplicity deemed wrong and complexity unsuitable, meaning that mental models are not to be too simplistic, whereas formal models must not be too complex. The Bonini-paradox (named after Professor Charles P. Bonini, Stanford Graduate School of Business), which describes the difficulties of designing complex system models and simulations, expresses a similar idea (for details refer to Bonini 1963). Starbuck (1976) explained this paradox in the 70s by stating that the more exhaustive the complex system models are, the less comprehensible they become. This means that the more realistic the models, the more difficult the task of understanding them. As a consequence, the real processes on which the models are based cannot be entirely understood. This fact makes it even more difficult to locate the decision-making dispositions. Summing up, the system dynamics formal modeling approach offers numerous advantages but bears some limitations nonetheless.

2.1.3 Advantages and Limitations

As is common for all other models, the system dynamics model is also subject to simplifications (Senge 2003). Therefore it is necessary to determine from the very beginning, which excerpt of reality will be examined for which purpose, so as to be able to identify the relevant and indispensible aspects, elements, interdependencies and forms of actions later on in the process of modeling. A subsequent model validation should be able to remove any uncertainty whether the model does not adequately reproduce the real system behavior in all its aspects (Bossel 2004).

The system dynamics approach can be applied in the fields of research, teaching and decision-making. These three fields are not entirely separate, but the boundaries between them can be blurred.

In the field of **research**, system dynamics models are based on physical laws or on problem-solving theories amended by empirical findings respectively (Nienhüser 1996; Sterman 2004; Ogata 2004; Weller 2007). Such models are precise, objective and verifiable (Strohhecker 2008), especially because of their symbolic and formal language, thus meeting the stringent quality requirements of science and research. Irrespective of this aspect, the quality of the prediction of the future is a major challenge for system dynamics as for any other simulation technique used as a scientific prediction method. Even the most elaborate models cannot guarantee entirely precise predictions (Eggers 1991). The theories and the assumed interdependencies on which the system dynamics model is based are valid until the derived hypotheses and findings are empirically disproved (Picot et al. 2008).

The system dynamics models can prove valuable in the field of **teaching** by elucidating causalities and interdependencies and making them comprehensible. Thus, such models are both operators and accelerators of teaching and learning processes (function of a catalyst), both concerning changes of behavior (single-loop learning: This expression was coined by Argyris (1985) and describes a learning process to reach current aims within the framework of existing mental models, i.e. interpret the information feedback with the help of the existing mental models (Sterman 2004).) and the adjustment of mental models (double-loop learning: This term was also coined by Argyris. Here, the information feedback of the real world does not only change decisions within existing decision-making rules, but also displays reciprocal effects and changes the underlying mental models (Sterman 1994; Moxnes 2000, 2004; Sterman 2004; Sterman and Sweeney 2007; Wheat 2007; Strohhecker 2008)).

In this context, it is especially the graphical representation of the interdependencies that facilitates the understanding of the system (Hafeez et al. 2004). Despite these advantages, it must be stated that system dynamics models are useful but not sufficient for better learning in themselves (Strohhecker 2008). During teaching and learning processes, these models and their simulations should be regarded as complementary to the teaching and learning methods already in use and not as their substitutes.

As experiments based on system dynamics models are supposed to lead to informed decisions and decision-making rules (Forrester 1961), they are often used in the field of **decision-making**. They can contribute to solving several problems (Strohhecker 2008) by summarizing fractions of the model to a holistic overall pattern, thus offering a more comprehensive representation of the decision-making problem (Hafeez et al. 2004; Strohhecker 2008). A wide range of possible action dispositions can be tested with a system dynamics model in order to find the best possible solution within the given interdependencies. This happens without performing trials on the original material (Bossel 2004), which would be partly impossible anyway. These decision-making experiments are consequently relatively cost and time efficient, bearing neither risks nor real consequences (Kramer and Neculau 1998). Even though the identification and representation of real and relevant interdependencies as well as the following creation of a model are time-consuming and can take, depending on the problem setting, a few weeks up to two years (Strohhecker 2008), it is assumed that experiments are more cost-intensive in the reality. At the same time, a system dynamics model can restore the initial system state, similar to a flight simulator for pilots (Strohhecker 2008).

In order to be reliable enough as an application for decision-making, the model should not contain any mathematical inaccuracies (Milling 1981). For the performance of a decision-making experiment with a system dynamics model, sufficiently precise quantitative data is required. Reference should be made to the fact that the data can often only be estimated or defined within ranges of values, which in turn causes (Coyle 2000) and increases (Hamilton 1980) for the quantification of qualitative variables the transfer of inaccuracies into the system.

2.1.4 Graphic Representations and Constitutive Elements

As briefly mentioned in Sect. 2.1.1 on page 10, different representations can generally be distinguished in general: on the one hand, the Causal-Loop-Diagrams (CLD), which enable a qualitative study of dynamic systems and on the other hand the Stock-Flow-Diagrams (SFD), which are used in quantitative studies.

The specific advantage of CLD is the relatively straightforward way of visualizing system structures and causalities as well as the prediction representation of expected qualitative system behavior. For a detailed quantitative study of dynamic systems, SFD will be used.

2.1.4.1 Causal-Loop-Diagrams (CLD)

CLD have long been a standard practice within the SD-approach. Today they are mostly used before the simulation analysis to aid in the representation of the basic causal mechanisms at work (Binder et al. 2004). CLD is therefore to be defined as a qualitative visualization of the interdependencies between different elements. Such a diagram consists of a set of nods which link the various elements or variables to each other. The relationships between the variables, visualized by means of arrows, can be labeled as positive or negative (Table 2.1).

CLDs help to identify whether the initial pulse. Here a pulse is not seen as a physical parameter, but more like in electrical engineering—a single, temporally limited and transient effect or a periodically repeated series of effects (refer to Rint 1953 and Schröder et al. 1972 for details). within a system will be reinforced or dampened. To determine which of the two will occur, the value of one variable X will, for example, be increased at one of the nods. At the other remaining nods, the variables which are influenced by the initial pulse are examined in regard to

Table 2.1 Definitions and examples for relationships in CLD

Symbol	Denomination	Meaning	Equation
X $\xrightarrow{+}$ Y	Positive causal relationship	X increases, Y increases as well or X decreases and Y decreases as well	$Y = \int_{t_0}^{t} (X + \ldots) ds + Y_{t_0} \frac{t}{dt}$
X $\xrightarrow{-}$ Y	Negative causal relationship	X increases, Y decreases or X decreases and Y increases	$Y = \int_{t_0}^{t} (-X + \ldots) ds + Y_{t_0} \frac{t}{dt}$

increases or decreases. A CLD is considered reinforcing or escalating when it retains its sign after the initial pulse and at the end of the simulation. If the final pulse is contrary to the initial one, this is an indicator that the system has a stabilizing effect. Thus, reinforcing or escalating systems can either be identified by their exclusively positive links or by an even number of negative links. Stabilizing systems, however, display an uneven number of negative links. It is expected that reinforcing systems result in an exponential increase or decrease, while stabilizing systems converge rather to a certain value. Consequently, determining whether a system is reinforcing or stabilizing is an important step for the prediction of dynamic system behavior at an early stage.

A simple example will demonstrate that a positive causal link is identified by two variables changing in the same direction. This means that an increase of the variable X at the initial nod directly causes an increase of the value of the dependent variable Y as well. On the other hand, a negative causal link means that both variables X and Y will change in opposite directions: if the X value increases at the initial nod, the dependent Y value will decrease and vice versa.

With regard to the causal relation between steel price, market supply and demand (Fig. 2.2), it can be observed that a rising demand causes higher prices, while a lower demand has a dampening effect. This is an example of a positive causal link, also named stabilizing loop system (balancing). A rising demand also increases the supply. The effect is time-lagged, represented in the figure by two parallel lines across the linking arrow. An increased supply, on the other hand, has a dampening effect on the steel price, representing a negative causal link or escalating loop system (reinforcing). Taken as a whole, the cycle demand-supply-steel-price is thus an escalating causal loop system.

This example serves the illustration of the difference between linear cause-and-effect chains and CLDs. In CLDs, the effects are also influencing their cause. Nevertheless, CLDs are neither comprehensive nor definitive, but should rather be considered tentative (Sterman 2004).

Fig. 2.2 Example of a causal loop diagram

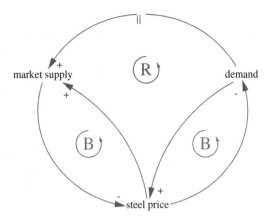

Table 2.2 Example of a
stock-flow-diagram

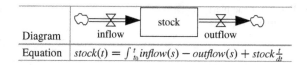

| Diagram | inflow | stock | outflow |

| Equation | $stock(t) = \int_{t_0}^{t} inflow(s) - outflow(s) + stock\frac{t}{dt}$ |

Using different scenarios, CLDs can depict causal links to elucidate the influence of specific factors on long-term trends, but these do not distinguish between stock variables and flow variables (Sterman 2004) and therefore are generally ill-suited for valid quantitative analyses. For the latter, stock-flow-diagrams are applied.

2.1.4.2 Stock-Flow-Diagrams (SFD)

As opposed to CLD, SFD distinguishes between stock variables (levels, stocks) and flow variables (flows, rates). Stocks are state variables, building the resource base like goods, machines or personnel, usually displayed in the SD-notation as rectangles. Stock variables change their values through flow variables (in- and outflow). The values of the stock variables are continuously changing over time, even when the share of flow variables, displayed in the SD notation as double line arrows, change discontinuously. The change intensity of such an in- or outflow is displayed as a valve (Table 2.2).

Hence, flow variables influence the values of stock variables. Contrary to stock variables, the value of a flow variable does not depend on prior values, but on its related stock variable in the system, where applicable together with exogenous factors like auxiliary or constant variables. It is especially important for the construction of SFDs to distinguish the stock and flow variables. A helpful tool is the virtual halt or freeze of a system. Here, stock variables still display values whereas flow variables are devoid of values. For further details refer to Sterman (2004). The set of all dynamic equations creates a system of non-linear differential equations simultaneously computing the change for each variable through integration across time in the relevant period (Niemeyer 1977).

An example from masonry will be used to illustrate the design of an SFD and the contained SD-elements on the one hand and to allow a short qualitative analysis on the other hand. A conscious decision has been made to omit an explicit description of the differential equations in this context. In the first example (Fig. 2.3), the stock variable is the *work to do*.

The stock variable *work to do* means here that a specific number of square meters of brickwork should be performed. This stock variable is defined by the *production* taking place, represented as flow variable, meaning that over time, the required amount decreases. In this example the stock variable should be regarded as a source out of which flows the "amount of brickwork". In this case it is irrelevant what happens with the produced good, displayed in the SD-notation as a cloud at the end of the production process. This symbol is therefore marking the system boundaries of the model. Here the production is being exogenously influenced by a number

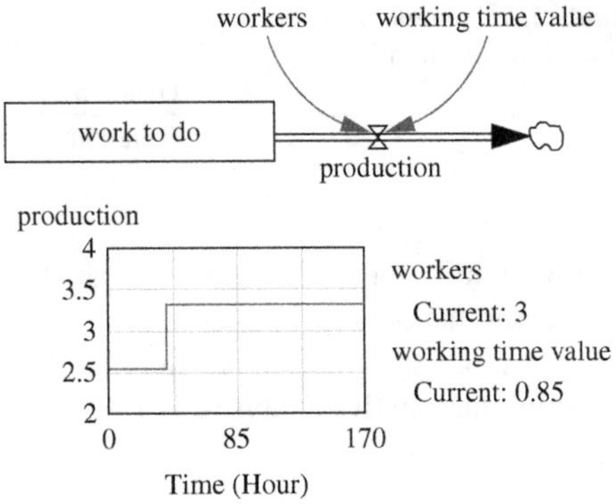

Fig. 2.3 Example 1: Wall construction work

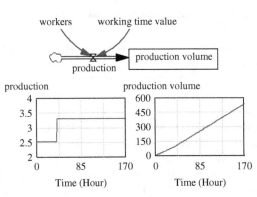

Fig. 2.4 Example 2: Production output wall construction

of *workers* with a specific productivity, displayed by the *working time value*. Both *workers* and the *working time value* are constant variables.

For the sake of the qualitative analysis, the *work to do* is set to the value of 450 m^2. The simulation time runs for 170 h. The number of *workers* is set to three and the *working time value* amounts to 0,85 m^2/h. The expected result of production is 2,55 m^2/h. But how much brickwork will have been completed at the end of the simulation? To answer this query, the stock variable *production* is to be examined. In general, the system boundaries at the end of *production* can also be set at the beginning and the stock variable at the end. In this case the stock variable is the *production volume* (Fig. 2.4).

The simulation demonstrates that with identical initial values 433,5 m^2 of brickwork will have been completed after 170 h. With regularly recurring tasks, an effect of adjustment to the job or *work experience effects* can reasonably be assumed. The extent to which the *production* is influenced by such effects should also be examined.

To this mean, the initially constant variable *working time value* will be upgraded and extended (Fig. 2.5). It is assumed for the given example, that after 40 h the productivity increases by 30 % and persists at this level for the rest of the simulation

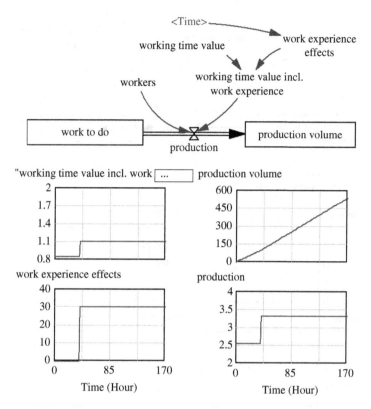

Fig. 2.5   Example 3: Example 2 including gained experience

period. Such an increase is a realistic value. Refereing to studies conducted by Ibbs (2012), actual productivity can exceed plan 60 % under the condition of no to few changes within the project.

The system boundary in the example is replaced at the beginning of the flow variable with the stock variable *work to do*. The diagrams clearly show that the *work experience effects* directly influence the variable *working time value incl. work experience* as well as the *production*. Nonetheless, the production level increases to the time t = 40 h only moderately. Interestingly, at time t = 170 h, a production level of approximately 533 m^2 has been performed. The legitimate question is thus, by which time the requested level of 450 m^2 will be reached. To this end, a termination criterion will be defined as shown in the final example (Fig. 2.6).

Apart from this, it is assumed that at the time t = 85 h additional 12 *workers* will be available. Along with the moderate rise in *production* because of the *work experience effects*, a steep increase of the production level occurs at time t = 85 h. Under the given conditions, 450 m^2 of brickwork are reached here in a time t = 98 h.

With the help of the examples, the elements existing within an SFD and their mechanisms of action ought to be introduced. Beyond that, the following should become comprehensible:

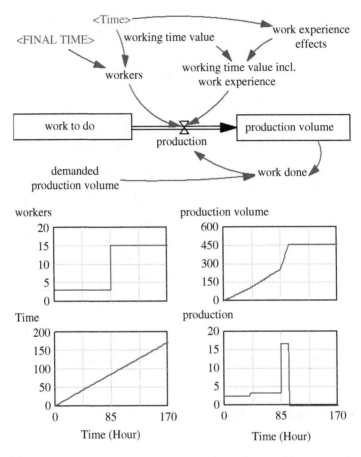

Fig. 2.6 Example 4: Example 3 including additional workforce and end of project

- The purpose of the simulation must be determined before the design of a simulation. (see example 1 & 2).
- It is advantageous to design a rough model first and to develop it gradually and subsequently (example 2 & 3).
- In order to identify the background and reasons for the occurrence of specific influences, the ceteris paribus clause is to be used in the qualitative analysis, meaning that only one element should be changed at a time (example 3 & 4).

Among others, it is these aspects which will be considered in Sect. 2.3.6, starting on page 31, in the preparation of a formal conceptual approach for modeling and simulating dynamic systems.

2.2 System Dynamics in the Field of Construction

The SD approach became particularly widespread in the 1980s in such fields as project management, shipbuilding, defense and aviation (Roberts 1978; Cooper 1980; Reichelt and Sterman 1990). Simulation is not just an academic exercise today. The second author of this text has used it extensively on numerous constructions to study a variety of issues; e.g. borrow pit locations and haul road routing; schedule duration analysis; and especially, loss of productivity disputes. He has worked with both owner and contractor clients to study delay and disruption disputes on matters such as rail systems, large canals, powerplant, and refinery projects. Therefore there are numerous SD-models in the various fields of stationary industry that are largely universally transferable. According to Bauer, the special feature of the construction industry is that it is a processing industry without own production site. The buildings are made in make-to-order production at a requested location (Bauer 2007) and various stakeholders are involved. Because of the one of a kind production in the non-stationary industry, the transfer of existing models from the stationary industry proves difficult. John D. Sterman also acknowledged this distinct feature of the construction industry in the 90s and related it to the SD approach. According to Sterman (1992), construction projects are highly complex and dynamic and consist of numerous interdependent elements, various feedback loops and non-linear relations and feature both "hard" and "soft" data.

Irrespective of the different research areas construction engineers are working on, a set of established methods and procedures can be observed in the construction practice. These methods and procedures are based on research findings and norms or have been transferred and adapted from other fields; the long-term effective ones form a code of practice readily available when needed. This fact should be taken into consideration when applying methods from other domains.

Although many researchers apply the SD-approach to problem-solving in the field of construction (e.g. Chang et al. 1991; Ogunlana et al. 1995; Ibbs and Liu 2005; Mbiti 2008; Mawdesley and Al-Jibouri 2010; Skribans 2010; Hou et al. 2011), there is currently no systematic approach to make the design of SD models practicable and applicable on the one hand, and to support the universal usability of construction-specific models on the other hand. That lack of systematic design is an important topic and one in which these two authors are interested and working on.

The obstacles frequently encountered when attempting to introduce or apply simulations in the field of construction, prove the necessity of such a systematic universal approach. An empirical study (Study SimBauDE: The current use of simulations in the German construction industry, carried out by the author in the year 2012) identifies these obstacles as lack of know-how, excessive costs and the increased effort associated with the application of the model. Hence, a readily applicable modeling approach would be helpful in both closing the gap between existing knowledge and the expertise required by the simulation and in reducing the modeling costs. Such an approach is not only justifiable by the mentioned obstacles it is supposed to overcome, but also by the fact that most test subjects would consider making increased use of simulations in their companies, extending the field of application of the approach beyond the usual field of research and development. Further, the approach appears especially promising in the management of organizational strategy and human resources. Here, the simulation is mainly regarded as a method to cut costs and minimize potential risks.

Despite the few basic elements it consists of (stock- and flow values, variables, auxiliary variables, constant variables), the SD approach seems to be seen in the literature as well as in its applications more as a "special method". The design of a specific method and of a feasible framework for the modeling and simulation of dynamic systems can contribute sensibly in this context, as it provides a purpose-oriented, useful and immediate access to system dynamics simulations within a domain.

2.3   Domain-Specific Libraries of Simulation Modules

Even though SD models are used in various fields of science to gain more insights, both theory and practice of theses simulations are rather difficult to comprehend for externals or third parties without previous SD knowledge. The SD approach may be well-known in many branches of science (Hübner-Dick 1980 for example, who researches possibilities and limitations of system dynamics in the analysis of international politics.), but for a comprehensive discussion—for instance to be able to develop valid and functioning models and simulations independently—both a detailed and thorough study of the theoretical relationships and an intensive search for previous models, which are potentially suitable for integration into the intended model, deems necessary. In addition to that, previously developed SD models quite often acquire their reputation within their specific area of a single domain only. Currently it appears to be inevitable that previously existing models (independent of their scientific subject) and their contained elements can hardly be integrated into new models. One significant reason can be the lack of a viable possibility to exchange existing SD models within a domain or even beyond it. This seems to point to the necessity to develop a general and feasible method to classify SD models as well as their contained components and units to make them available for other modelers. One possibility is the development of domain-specific libraries of simulation models, called system dynamics libraries in the following.

2.3.1   System Dynamics Libraries

Following the nomenclature of chemistry, a system dynamics library (SDL) is based on an atom-molecule-component-approach (abbreviated: AMCA). Therefore such a library consists of the following three fundamental entities (E): Atom (a), Molecule (m) and Component (c). In contrast to the nomenclature of chemistry the term molecular substance is not used here.

An atom is the smallest entity (Nic et al. 2006) which can still be characterized in a model. To these units belong all single entities which exist independently and without external influence inside a model. Following general model theory, atoms have defined attributes and properties (Stachowiak 1973). Atoms cannot cause system changes by themselves only, but may do so in combination with other entities existing in a model. Consequently, all discrete entities are atoms, e.g. stock and flow variables, (auxiliary) variables and constants. A coupling of atoms forms a molecule (Nic et al. 2006), which gains its properties through the interaction of its atoms. The linking of molecules results in a component or a module which has a case-sensitive internal processing logic. The combination of atoms, molecules and components results in a comprehensive model.

As the application of system dynamics models is supposed to lead to more consolidated decisions and decision-making rules (Forrester 1961), the decision-making level (D) is included as well in the approach introduced here. Therefore the operative, the tactical and the strategic decision-making level are equally comprised. The operative level contains mostly the physical realization and the implementation of tasks. On the tactical level, the processes and the organization within a corporation play a central role whereas the position and the targets of the corporation in the market are central focus of the strategic level.

To derive effective decision-making rules for various functional areas (F) within a company with the help of a holistic system analysis of complex dynamic problem settings, it appears sensible to include the following areas: strategy and organization (SO), research and development (RD), finance and governance (FG), marketing and sales (MS), human resources and leadership (HL), as well as operations and procurement (OP). The structure of the functional areas follows van Assen et al. (2011).

With the help of such a system (Fig. 2.7), different SDL can be developed on the basis of the SD approach. These libraries offer a helpful transfer potential for their own domain, but also make units available beyond their own domain and allow the integration in their branch of science with little or no adaptations. This classification of units creates a high degree of universality and reusability. Previously developed entities for a specific decision-making level or functional level can be used in other levels with potentially little or no adaptations.

Using this approach does not only mean that new entities in development can be classified systematically, but existing models (independent of the domain they were developed for) can be analyzed methodically, the separate entities can be extracted and made available in a specific SDL. Thus, an SDL can be continuously complemented by controlled and systematic examination of SD models.

Atoms and molecules represent the universal entities in an SDL, whereas components were already assigned a fixed point of reference within a precise problem setting in a specific level. The boundaries between the different levels can be blurred, depending on the selected system limitations and the desired scale. If, for example, a company is modeled as a whole on the strategic level, considering details or individual processes of the work flow deem inappropriate (Troitzsch 2004). In this case a molecule or a component, which was developed in detail for the operative

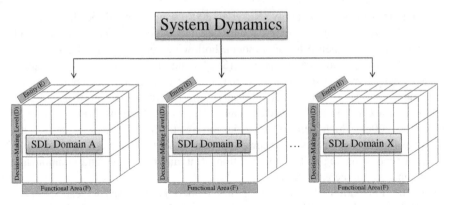

Fig. 2.7 Concept of system dynamics libraries (SDL)

level, can be reduced to an atom in a model of the strategic level. Here, the attribute required for the new model will be derived as a time-dependent function from the previous entity. If the need to inspect the behavior of this atom in more detail is identified later on, it can effortlessly be expanded to the previous stage.

2.3.2   Synergistic Modeling and Simulation Using SDL

As the SD approach comprises both the decision-making level and the functional area of a company, this approach offers a formal basis for the synergistic modeling and simulation within a domain. Consequently, the consistent development of a company-specific SDL leads to numerous possibilities of analysis (Fig. 2.8).

With the help of already available reference models adapted to the individual processes, activities and data of a company within a necessary customizing process, the division managers, e.g., of functional areas, are put in the position to analyze their area of responsibility in due consideration of several other levels of decision-making in the company. The inclusion of SDL units from more than only one functional or decision-making level can lead to (a) the development of more informed decisions and choices, which consider interests of other decision-makers in the company, (b) new insights, which can illustrate the previously unknown far-reaching scale of an evaluated decision, (c) the detection of potential conflicts which can be counteracted in advance and (d) the development of argumentation to convince individuals both in the own functional area but also from other involved decision-making levels that the targeted decision is sensible and appropriate. However, these exemplary possibilities of application depend on an SDL which is properly maintained by all departments in the company.

Regardless of the simplicity of this approach, it must be noted that specific competencies are needed for modeling and simulation. Additionally to the domain-specific skills the user needs knowledge of various models and simulations and when

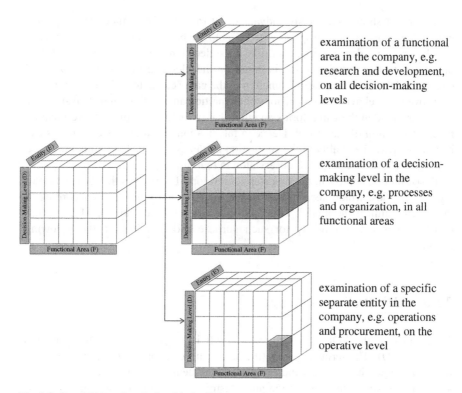

examination of a functional area in the company, e.g. research and development, on all decision-making levels

examination of a decision-making level in the company, e.g. processes and organization, in all functional areas

examination of a specific separate entity in the company, e.g. operations and procurement, on the operative level

Fig. 2.8 Possibilities of analysis with the SDL approach

it is beneficial to integrating them. It is essential to have the ability to understand the appropriate uses and limitations of different models. Further, users need technical skills to integrate those models and the ability to interpret the output and apply it in a meaningful way to find answers for previously defined questions.

In addition to the possibilities mentioned above, such a synergistic modeling approach will lead to further substantial benefits. As already mentioned in Sect. 2.1 on page 13, a model is only valid until the gained hypotheses and insights are disproven on empirical basis. This means that the gained results do not necessarily have to be true, whether due to analytical, formal or fact-based logic relationships. Both the units deemed relevant for the model and the links among these are subject to assumptions of at least one individual. Therefore it is true for such models as well, that wrong premises like units and/or erroneous deductions in form of links between the units may imply wrong conclusions. Strictly spoken, a model can only be accepted as true after it was definitely verified. Usually this would suggest an empirical study in form of a long-term observation of real relationships and behaviors, the results of which can be juxtaposed to the results of the simulation afterwards (further details regarding these aspects are discussed in Karl 2014).

This point shows the decisive advantage of the synergistic modeling approach with which the model designer will be able to build a feasible link between simulation and a decision game. If, ideally, a decision game has been established in parallel to the simulation and is based on the identical model (e.g. developed out of an SDL), the "time lapse" function of the game can help to form an empirical basis from the behaviors and decisions of the participants. A thorough analysis of the decisions made in the game forms an empirical basis to assist in a first verification resp. falsification of the model. Hence, a falsification based on these premises can help to improve the quality of the simulation model as a whole but without the need to monitor the whole period of observation in real time.

Based on this approach, the first step to develop SDL units is the localization of relevant entities, which need to be formally and explicitly described according to the specific aim of the model. This description is achieved with the SDL process ontology, devised specifically for such purpose and described in the following chapter.

2.3.3 SDL Process Ontology

Each entity of the SDL is structured in itself in a specifically laid-out process ontology (PO). The term ontology (the study and categories of being) originates from philosophy but is, in computer sciences, frequently seen as a clearly separated and formally structured description and illustration of terms, components etc. and their relation to each other in a given area of interest. Such ontologies are often employed for the organized exchange of knowledge (e.g. knowledge representation in the section of artificial intelligence). In contrast to taxonomies, which only display a hierarchical sub-categorization only, ontologies are in the position to depict relationships between the individual terms. For further details, please refer to, Uschold and Grüninger (1996), Oberle et al. (2009), among others. In the first stage, the SDL PO lists the resources included in a specific process. These are subdivided in, e.g., material, tools (e.g. equipment or machinery), personnel and capital as well as the dependencies between the individual resources (Fig. 2.9).

This means that atoms, for instance, can be allocated to the monitored resources. These resources contain the properties allotted to the atoms which, in turn, results in a dependency on the according functional areas. The second stage SDL PO integrates the previously defined objects of the first stage within a larger context and therefore creates a relationship between the individual processes. As a consequence, the previously introduced molecules consider the resources and properties of the atoms in a larger scale and allow the illustration of processes. These represent cost- and production-oriented activities and are summarized as process chains in the components. Hence, the application of a PO predefines the first inner structure of a domain-specific SDL, thereby allowing model developers to quickly locate the required elements for the respective model purpose and include them in their individual models.

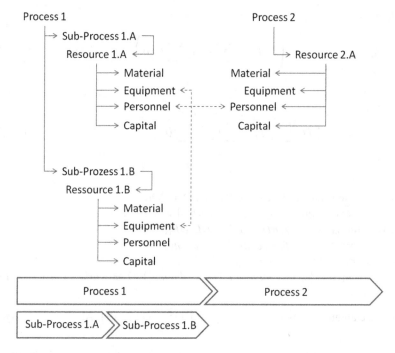

Fig. 2.9 SDL process ontology

On the basis of the SDL PO, first formally distinct and transferable networks of processes, objects and interdependencies are designed and can be drafted in more detail with the help of a specific SDL notation.

2.3.4 SDL Notation

Fundamentally important for the systematic expansion of an already developed SDL, an according notation is as elementary as the structured basis supplied by the PO. The notation aims at the possibility to establish SDL as database-supported libraries in the long term. This notation accounts especially for the formal predicate logic. To facilitate the denominations, each entity is assigned a unique identification number (uid) within its own field. With the assistance of further attached uids, the development respectively the affiliation of an entity within a group can be understood in the future. The separate units are defined as follows:

Atom:

$$a_{R,,uidn.\,...\,uidm} := a\left(D, p, F(p)\right) \in f(p)\,[u(p)] \tag{2.1}$$

Molecule:

$$m_{P,uidn.\,...\,uidm} := m\left(D, F(p), \sum_k (a_k), \sum_l (m_l)\right) \in f(p)\,[u(p)] \tag{2.2}$$

Component:

$$c_{P,uidn.\,...\,uidm} := c\left(D, \sum_j (Fp_j), \sum_k (a_k), \sum_l (m_l), \sum_m (c_m)\right) \in f(p)\,[\sum_j (up_j)]$$
$$\tag{2.3}$$

mit:

D = decision-making level $\in [CMD, CCD, CPD]$;
p = predicate $\in [property\ 1, \ldots, property\ i]$ with i$\in \mathbb{N}$;
F(p) = function $\in [SO, RD, FG, MS, HL, OP]$;
f(p) = form $\in [stock, flow, const, var, system]$;
u(p) = unit $\in [SI - unit, currency - unit, \ldots]$;
R = resource $\in [MA = material, DE = device, WO = worker, CA = capital, ST = storage]$;

(In case an atom is used as auxiliary variable in the model, AUX = auxiliary is used)

P = process $\in [PL = planning, SF = site\ facilities, EA = earthworks, SC = shell\ construction]$;
uidn = unique identification number n $\in \mathbb{N}$;
uidm = unique identification number m $\in \mathbb{N}$;
k = uid of included atom;
l = uid of included molecule;
j = uid of included component;

It is necessary to keep in mind, that a predicate p can be a conglomerate of different properties, e.g. the resource material can contain a quality, a weight and a price (arity of three). Hence, if an atom consists of a predicate with an arity of x and is subject to an itemization process in modeling, it needs to be broken down to further independent atoms with an arity of one to allow its use in the following simulation. Within a simulation, an atom is always required to be downgraded until it describes only one property. According to the first example in Sect. 2.1.4.2 on page 18, the notation is demonstrated as follows:

$$worker:\ a_{WO,1} := a\,(CPD, number, OP) \in const\,[qty.] \tag{2.4}$$

$$working\ time\ value:\ a_{WO,2} := a\,(CPD, working\ time\ value, OP) \in const\,[m^2/h] \tag{2.5}$$

$$work\ to\ do:\ a_{MA,1} := a\,(CPD, quantity\ of\ work, OP) \in stock\,[m^2] \tag{2.6}$$

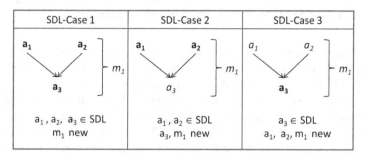

Fig. 2.10 SDL model cases

$$production : a_{MA,2} := a\,(CPD,\, quantity\ of\ production,\, OP) \in flow\ [m^2/h] \tag{2.7}$$

$$production : m_{SC,1} := m\,(CPD,\, OP,\, a_{WO,1} a_{,WO,2}) \in flow\ [m^2/h] \tag{2.8}$$

As can be seen, the flow variable *production* can appear as molecule or as an atom in an SDL. Indeed, the entity *production* is dependent on the atoms *worker* and *working time value* and consequently represents a molecule. Equally correct is the fact that this flow variable—on closer and separate inspection—qualifies as one smallest identifiable entity as it can neither cause any system changes on its own, nor exists the necessity to influence exogenously.

This described situation is only based on the fact that the modeling of entities using the SDL approach may need to consider the occurrences of feedback. Three cases can be distinguished in general (Fig. 2.10). In the first case, the bottom-up method (abstraction in increments following VDI 3633 (1996)) helps to form a molecule out of three already existing atoms. Here, two atoms influence the third one (e.g. var + const → var, var + const → flow, var + const → stock). The second case presupposes the existence of at least two more atoms, of whose coaction a previously unknown third atom is created. This means that one further atom was created during the construction of a molecule. The third case represents a recursive form, in which the implementation of the top-down-method (itemization in increments following VDI 3633 (1996)) transforms an existing atom into a molecule. For this, possible influences on the atom were identified. The result of this observation is the creation of more atoms.

As models on the basis of an SDL can be of substantial size and complexity—depending on the problem setting and the desired insights—a unified graphic mode of representation is introduced in the following. It can be used to facilitate the observer's understanding of complex models and elucidate the main features of an SDL-based model.

2.3.5 Unified System Dynamics Modeling Language

Although the SD approach provides two generally applicable and straightforward comprehensible modes of presentation to the user (CLD, refer to p. 15 and SFD, refer to p. 17), the introduction of a distinct modeling language appears sensible both for the support of the systematic expansion of SDL and for the improved understanding of the complex models and their underlying system. This modeling language is based on the concept of the already existing Unified Modeling Language (UML: A graphic modeling language for the specification, construction and documentation of software elements and other systems.) and adopts especially the presentation modes of the UML object, class and component diagrams. Thus, a specific symbol is assigned to each of these three basic units of an SDL (atom, molecule, component) which conveys information to the observer (Fig. 2.11).

Based on the SDL notation, the Unified System Dynamics Modeling Language (USDML) offers both a simplified and an elaborate presentation. The elaborate and more detailed presentation displays seven information units: 1. decision-making level (D), 2. functional area (F), 3. resource (R) respectively process (P), 4. property (p), 5. form (f), 6. unit (u) and 7. unique identification number (uid). In contrast to that, the simplified presentation consists of four necessary information units: 1. decision-making level (D), 2. functional area (F), 3. property (p) and 4. unique identification number (uid).

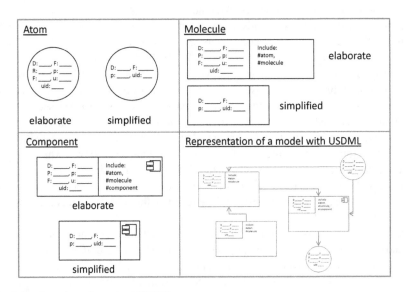

Fig. 2.11 Overview of notation USDML

2.3.6 Modeling and Simulation with the SDL

Here, the top-down method is dominantly applied in the model creation processes. The advantage of this approach, seen in contrast to the bottom-up method, is the applicability of the top-down method both in the modeling process itself and in the system analysis. In the modeling procedure, this preferred method is used to refine sections of the model in accordance to the specific problem setting. At first, some sections are modeled on a high level of abstraction, in which selected subsections are treated as black boxes (A model in system theory in which a unit-step response processes the input of a system and returns the feedback of the system on its environment (output), more details ref. to VDI 3633 (1996)). After that, each black box is subject to an individual and detailed modeling. Within these generated subsystems (consisting of at least one mathematical function), simulations are conducted and their results can either be seen individually or their input/output behavior can be used as basis for the superordinate model (application of a bottom-up method ref. to VDI 3633 (1996) for an incremental summary of results gained from the black boxes.).

To prepare the following examinations, a development framework is introduced for the modeling and simulation of dynamic systems; it includes the SDL approach (on basis of VDI 3633 (1996), Bossel (2004) and Sterman (2004)). The methodology we use here has the advantages that it is easy to understand and to implement. The user recognizes immediately what is necessary in the different development steps, what he has to do and especially why it is needed. In Addition, every step is oriented towards a specific output, so the user is always focused on a particular result for each development step.

In general, the modeling and the subsequent simulation including the presentation of the results and their discussion is formally separated in six phases. In this classification, the modeling comprises (1) identification and formulation of problem setting, (2) compilation of effect relationships and (3) differentiation of model concept. In the area of simulation, the three remaining phases can be found: (4) preparation and implementation of simulation, (5) presentation of results and lastly (6) discussion of results. Figure 2.12 depicts this development framework. It contains reasonable steps of the overall procedure, recommendations regarding their implementation as well as the preferred and intended results of the phases.

According to the development framework, the hitherto developed model should be transferred into suitable simulation software in phase four. To address this aim, the following chapter is concerned with the aspect of the software selection for the SD simulation. First, the current state of available computer programs is considered, then one software product is selected and used in the subsequent procedures.

Development framework for modeling and simulation based on the SDL approach

Phase	Step of procedure	Why is it necessary?	What is to be done?	Output
1. Identification of problem setting	definition of model objective (what, which motivation), formulation of written model , system building (black boxes)	specification of targeted knowledge gain, identification of substantial system variables and effect relationships, definition of system limits	written description of model and first reduction of reality	text models
2. Determination of effect relationships	block diagram as sum of all black boxes, identification of first relevant effect relationships (system variables and effects/influences)	illustration of first draft of system structure, derive statements on potential behavior, execution of qualitative research	derivation of written model description (only direct effects, isolated consideration, ceteris paribus condition), numerical inspection, mathematical analysis, effect matrix, Vester's "paper computer"	first mental models as block diagrams, effect relationship diagrams, e.g. causal loop diagrams or others
3. Differentiation of model concept	SDL PO for the identification of essential key variables, parameters (fixed system parameters), state variables (storage variables), modifier (interim/auxiliary variables)	generation of context and relationships between the individual processes within a model	successive differentiation of previously developed mental system and structuring in processes, sub-processes and involved resources	tables, allocation network etc.
	development/ extention of an SDL 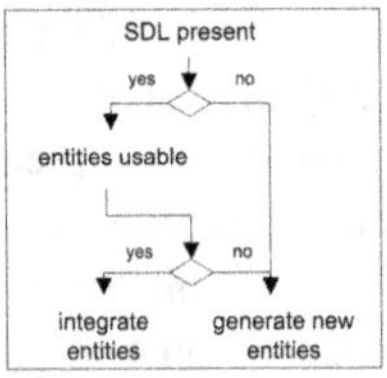 SDL present yes no entities usable yes no integrate entities generate new entities	long-term minimization of development efforts by re-utilization of already developed (possibly from other domains) units, first direct access to subsequent SD simulation, facilitated integration of new entities in different functional areas and decision-taking levels, basis for a computer model which can be simulated	first formal illustration of variables in the model (according to SDL PO defined classification), recursive or iterative formulation of variables following AMC approach	first general equations based on the SDL notation, grouped according to SDL PO, decision level and functional area
	USDML presentation of the general model	identification of relationships and missing units	summary of SDL entities into overall and general model	USDML diagram
4. Sim. preparation/ implementation	determination and presentation of differential equations, validation and model validity respectively check of system stability/ integrity	acquisition of reliable qualitative results, testing of model validity, potentially optimization of model	collection of valid and reliable data, transfer model into an appropriate simulation software (potential viewing of partial models and subsequent linking)	mathematical equations and simulation results
5. presentation of results	Evaluation of simulation results for gaining insights	comparison of simulation results to original model purpose	graphical preparation and/or generation of animations	tables, charts, diagrams, animations and other forms
6. Discussion of results	identification of significant results	specification of result effects on a) model purpose, b) other questions	discussion of the significant results and connecting to central question	text and graphical representations, if necessary

Fig. 2.12 Modeling and simulation with SDL

2.3.7   Software Selection for the SD Simulation

When choosing the most appropriate modeling software, the difference of qualitative or quantitative perspective needs to be considered. A qualitative view focuses exclusively the cause-effect relationships. Here, the system structure, the relationships and the influences of individual factors on each other are of dominant interest. The system behavior can be analyzed, but numerically-oriented predictions can only be achieved with the help of a quantitative perspective.

First it has to be decided whether the qualitative model design should be followed by a simulation, which attempts to gain insights and result in predictions using concrete (real or realistic) data. If this is not the case, all software programs which can generate texts and at least rudimentary graphics could in principal be used to display models.

The authors inspected and used a lot of software solutions. Due to the target group for which the following models should be developed (research, education and industry), emphasis is placed on the identification of an efficient, competitive and universally applicable software without any limitations. Independent of offered version with time limitations or other restrictions, the programs Mapsim, NetLOGO, Sphinx SD Tools, SystemDynamics and Vensim are eligible for the desired purposes. To guarantee a long-term availability and usability of the models, it has to be ensured that the selected software is sufficiently supported and can also be expanded with additional functions to carry out further analyses and simulations on the basis of the developed models.

Of these above-mentioned software products, especially the program Vensim appears to be suitable. Developers and suppliers of Vensim, Ventana Systems, Inc., work with SD-based simulations since the middle of the 1980s and can currently be seen as established in the market. The PLE version of Vensim (personal learning edition) is fully functional and at no charge for the personal use respectively for research and teaching. Furthermore, it is equally possible to export models developed with Vensim to other programs, e.g., Forio Simulations. Beyond this, models from other programs, e.g. from Stella and ithink, can be imported by Vensim with the help of an interface program. The functions contained in the PLE version of Vensim are suitable for basic or complex system dynamic models. A future expansion of functions like, e.g. Sensitivity Simulations with the Monte Carlo method and Live Data Connections can already be integrated with the PLE Plus version. Further functions are offered by the version Vensim Professional and Vensim DSS (Decision Support System, for details about DSS please refer to, e.g. Sauter 2010 and Turban 2011). The latter offers a development tool which helps to construct so-called management flight simulators. Subsequently, these can be used as independent applications. In all above-mentioned version of the software, the models created with the PLE version can be imported and integrated. Therefore, the PLE version will be used both for the following modeling as well as the simulation later on.

References

Argyris C (1985) Strategy, Change, and Defensive Routines. Pitman, Boston

Assen van M et al. (2011) Key management models. Prentice Hall Financial Times, Harlow

Baron J (2000) Thinking and deciding, 3rd edn. Cambridge Univ. Press, Cambridge

Bauer H (2007) Baubetrieb, 3rd edn. Springer, Berlin, Heidelberg

Binder T et al. (2004) 'Developing system dynamics models from causal loop diagrams', vol 22, pp 1–21

Bonini CP (1963) Simulation of information and decision systems in the rm. Prentice-Hall, Englewood Clis

Bossel H (2004) Systeme, Dynamik, Simulation: Modellbildung, Analyse und Simulation komplexer Systeme, 1st edn. Books on Demand GmbH, Norderstedt

Chang CL et al. (1991) 'Construction project management: a system dynamics approach', vol 9, pp 108–115

Cooper KG (1980) 'Naval ship procurement: a claim settled and a framework built'. Interfaces 10(6):20–36

Coyle G (2000) 'Qualitative and quantitative modelling in system dynamics: Some research questions'. Syst Dyn Rev 16(3):225–244

Doyle JC et al. (1992) Feedback control theory. Dover Publications, New York

Doyle JK, Ford DN (1999) 'Mental models concepts revisited: some clarications and a reply to Lane'. Syst Dyn Rev 15(4):411–415

Eckey HF et al. (2011) Ökonometrie: Grundlagen - Methoden - Beispiele, 4th edn. Gabler, Wiesbaden

Eggers B (1991) 'Szenario-Technik'. Das Wirtschaftsstudium 20(10):705–705

Forrester JW (1961) Industrial dynamics. MIT Press, Cambridge

Forrester JW (1968) Principles of systems, 2nd edn. MIT Press, Cambridge

Forrester JW (1969) Urban dynamics. MIT Press, Cambridge

Forrester JW (1971a) 'Counterintuitive behavior of social systems'. Technol Rev 73(3):52–68

Forrester JW (1971b) World dynamics. Wright-Allen Press, Cambridge

Forrester JW (1972a) Grundzüge einer Systemtheorie. Gabler, Wiesbaden

Forrester JW (1972b) Industrial dynamics, 7th edn. MIT Press, Cambridge

Gold S (2005) 'System-dynamics-based modeling of business simulation algorithms'. Simul Gaming 36(2):203–218

Hafeez K et al. (2004) 'Human resource modelling using system dynamics', vol 22, pp 120

Hamilton MS (1980) Estimating lengths and orders of delays in System Dynamics models, pp 162–183. Elements of the system dynamics method. MIT Press, Cambridge

Hübner-Dick G (1980) Simulation internationaler Beziehungen, Möglichkeiten und Grenzen von 'System Dynamics' in der Analyse internationaler Politik. Haag + Herchen, Frankfurt/Main

Hou W et al. (2011) 'Payment problems, cash flow and protability of construction project: a system dynamics model', pp 693–699

Ibbs W (2012) 'Construction change: likelihood, severity, and impact on productivity'. J Leg Aff Dispute Res Eng Constr 4(3):67–73

Ibbs W, Liu M (2005) 'System dynamic modeling of delay and disruption claims'. Cost Eng 47:12–15

Johnson-Laird PN (1983) Mental models: towards a cognitive science of language, inference, and consciousness. Harvard University Press, Cambridge

Karl CK (2014) 'Solving the Simulation Paradox - How Educational Games can support Research Efforts'. In Developments in Business Simulation & Experiential Learning, vol 41, p 132–139, Orlando, Fl. Association for Business Simulation and Experiential Learning

Kramer U, Neculau M (1998) Simulationstechnik. Hanser, München

Mawdesley MJ, Al-Jibouri S (2010) 'Modelling construction project productivity using systems dynamics approach'. Int J Product Perform Manag 59:18–36

Mbiti TKP (2008) 'A system dynamics model of construction output in Kenya'. Master's thesis

Milling P (1981) Systemtheoretische Grundlagen zur Planung der Unternehmenspolitik. Duncker+Humblot, Berlin

Milling P (1996) Simulationen in der Produktion, pp 1839–1851. Handwörterbuch der Produktionswirtschaft, 2nd edn. Schäffer-Poeschel, Stuttgart

Moxnes E (2000) 'Not only the tragedy of the commons: Misperceptions of feedback and policies for sustainable development'. Syst Dyn Rev 16(4):325–348

Moxnes E (2004) 'Misperceptions of basic dynamics: The case of renewable resource management'. Syst Dyn Rev 20(2):139–162

Nic M et al. (2006) International union of pure and applied chemistry "Gold Book". ICT Press, Prague

Niemeyer G (1977) Kybernetische system- und modelltheorie: system dynamics. Vahlen, München

Nienhüser W (1996) Die Entwicklung theoretischer Modelle als Beitrag zur Fundierung der Personalwirtschaftslehre. Überlegungen am Beispiel der Erklärung des Zustandekommens von Personalstrategien, pp 39–88. Grundlagen der Personalwirtschaft: Theorien und Konzepte. Gabler, Wiesbaden

Oberle D et al. (2009) What is an ontology? pp 1–17. Handbook on ontologies, 2nd edn. Springer, New York

Ogata K (2004) System dynamics, 4th edn. Pearson Prentice Hall, Upper Saddle River

Ogunlana S et al. (1995) 'Civil engineering design management using a dynamic model', vol 13, pp 757–765

Picot A et al. (2008) Organisation: Eine ökonomische Perspektive, 5th edn. Schäffer-Poeschel, Stuttgart

Reichelt KS, Sterman JD (1990) Halter marine: a case study of the dangers of litigation. Massachusetts Institute of Technology, Sloan School of Management, Cambridge

Richardson GP (1991) System dynamics: simulation for policy analysis from a feedback perspective, pp 144–169. Qualitative simulation modeling and analysis. Springer, New York

Rint C (1953) Handbuch für Hochfrequenz- und Elektrotechniker, vol. 2. Bd. Verlag für Radio-Foto-Kinotechnik GmbH, Berlin-Borsigwalde

Roberts EB (1978) Managerial applications of system dynamics. Productivity Press, Cambridge

Sauter VL (2010) Decision support systems for business intelligence, 2nd edn. Wiley, Hoboken

Schröder H et al. (1972) Elektrische Nachrichtentechnik, vol 3. Bd. Verlag für Radio-Foto-Kinotechnik GmbH, Berlin-Borsigwalde

Schwarz R, Ewaldt JW (2002) Über den Beitrag systemdynamischer Modellierung zur Abschätzung technologischer Evolution, pp 159–175. Technologie-Roadmapping, Zukunftsstrategien für Technologieunternehmen. Springer, Berlin, Heidelberg

Senge PM (2003) Die fünfte Disziplin: Kunst und Praxis der lernenden organisation, 9th edn. Klett-Cotta, Stuttgart

Skribans V (2010) 'Construction industry forecasting system dynamic model'. In: Proceedings of the 28th international conference of the system dynamics society, vol 28, pp 1–12

Stachowiak H (1973) Allgemeine modelltheorie. Springer, Wien

Starbuck WH (1976) Organizations and their environments, pp 1069–1123. Handbook of industrial and organizational psychology. Rand McNally College, Chicago

Sterman JD (1992) 'System dynamics modeling for project management'. Tech. rep., Massachusetts Institute of Technology, Sloan School of Management

Sterman JD (1994) 'Learning in and about complex systems'. Syst Dyn Rev 10(2–3):291–330

Sterman JD (2000) Business dynamics: systems thinking and modeling for a complex world. McGraw-Hill, Boston

Sterman JD (2004) Business dynamics: systems thinking and modeling for a complex world. McGraw-Hill, Boston

Sterman JD, Sweeney LB (2007) 'Thinking about systems: Student and teacher conceptions of natural and social systems'. Syst Dyn Rev 23(2–3):285–311

Strohhecker J (2008) System Dynamics als Managementinstrument, pp 17–33. System Dynamics für die Finanzindustrie. Simulieren und analysieren dynamisch-komplexer Probleme, 1st edn. Frankfurt-School-Verlag, Frankfurt am Main

Troitzsch KG (2004) 'Simulationsverfahren'. Das Wirtschaftsstudium 33(10):1256–1268
Turban E (2011) Decision support and business intelligence systems, 9th edn. Pearson, Boston
Uschold M, Grüninger M (1996) 'Ontologies. Principles methods and applications'. Knowl Eng
    Rev 11(2):93–155
Valéry P (1937) Notre destin et les lettres - Regards sur le monde actuel et autres essais. Gallimard,
    Paris
VDI 3633 (1996) 'Simulation von Logistik-, Materialuss- und Produktionssystemen - Begrisde
    nitionen (Entwurf)'. Beuth Verlag, Düsseldorf
Weber L, Schwarz D (2007) Ergebnisse eines Methodenvergleichs - Prognosefehler und quantita-
    tive Aussagen zur Personalstruktur von Unternehmen in verschiedenen Regionen, pp 193–208.
    Alterung im Raum. Auswirkungen der Bevölkerungsalterung unter besonderer Berücksichti-
    gung regionaler Aspekte. Books on Demand, Norderstedt
Weller I (2007) Fluktuationsmodelle: Ereignisanalysen mit dem Sozio-oekonomischen Panel,
    vol 28, 1st edn. Hampp, München
Wheat D (2007) 'The feedback method of teaching macroeconomics: Is it eective?' Syst Dyn Rev
    23(4):391–413

Chapter 3
Exemplary Development of a Model Library

Apart from the concept of a domain-specific SDL, which is based on the SD approach and called construction dynamic library (CDL, Fig. 3.1), a cadre of module-oriented models will be introduced. Within the CDL, three significant areas are in the central focus: (1) the construction project: Construction Project Dynamics (CPD) (2) the construction company: Construction Company Dynamics (CCD) and (3) the construction market: Construction Market Dynamics (CMD). The CDL is supposed to enable users to depict both the construction industry and all involved parties in a network of linked operational, economic and market-dynamic processes.

These models and the included units are to offer the basis for further developments and should be summarized using the term construction dynamics (CD). The design of the different units will be shown in detail in the following to demonstrate the practical development of the CDL, taking the operative level (CPD) as an example. As the modeling of CD units is a recursive process (refer to Fig. 2.12 on page 32), the development of the exemplary units is also intentionally illustrated in this form. As a consequence, the potentially missing atoms needed in Sect. 3.6 on page 45 and the equally lacking units in Sect. 3.7 on page 57, respectively, are going to be modeled within these chapters. As the formal approach is analogue both on the tactical as well as the strategic level, an explicit illustration of the development process on these levels is deliberately omitted.

3.1 Purpose of the Model

A project model is supposed to be generated on the operative level while applying the SDL approach. The aim is to have a project model which can display connections and dependencies or predict, both correctly and qualitatively, tendencies and dynamics of developments. These predictions should be formed on the basis of,

© The Author(s) 2016
C.K. Karl, W. Ibbs, *Developing Modular-Oriented Simulation Models Using System Dynamics Libraries*, SpringerBriefs in Electrical and Computer Engineering,
DOI 10.1007/978-3-319-33169-0_3

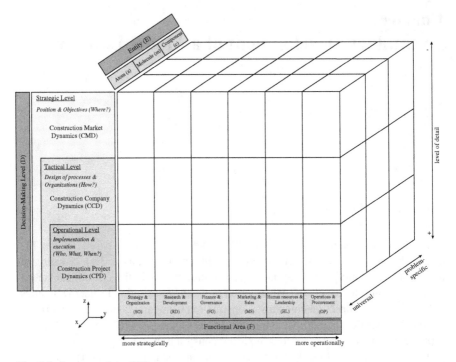

Fig. 3.1 Structure of the CDL

e.g., risks in the construction operation itself, volatilities of labor or operating costs
as well as the potential influence of experience and necessary on-the-job training
required by the personnel of a construction site in a surface construction project.
Additionally, elementary supply chains will be shown in the project model to allow
examination of their influence on the project. Primarily intended orientation of the
model *construction project* are the involved costs. The following questions should
be answered with the help of such a model and the subsequent simulation:

- How do the costs change with the passing of time?
- Determination of the optimal warehousing and quantity stored at minimal costs.
- Does warehousing pay off? If yes, which type and in which quantities?
- How much storage capacity is necessary for this?
- Which influence exert fluctuations on the bulk flow (choke points or similar)?
- How critical is price volatility during a construction project?
- In how far do workforce-specific parameters like experience and training-on-the-
  job influence the production?
- What influence do specific risks have and how do the effects change over time?

3.2 System Limits

Only surface construction is targeted in the model *construction project*. Furthermore, only the operations from the setup of the construction site to the completion of the building shell are considered in detail. These, in turn, are defined by the primary processes earthwork, reinforced concrete construction including formwork, reinforcements and concrete works, as well as brickwork. Resources like material, tools and personnel are included in the calculations. The remaining unconsidered tasks are seen as services contributed by subcontractors or third-parties.

3.3 Fundamental Cause-Effect Relationships

First, the model *construction project* is conceptualized on the basis of the previously defined system limitations and then, in the following chapters, qualitatively modeled in detail using the top-down-method (VDI 3633 1996). At this stage, the model *construction project* is built on fundamental reference figures and their relationships to each other (Fig. 3.2). The reference figures needed later for the simulation can be acquired from real projects or from sources in literature (e.g. statistical databases for construction costs, in Germany (Baukosteninformationszentrum 2008)). A first project definition is determined with the following input parameters: (a) gross volume (GV), (b) gross floor space (GFS), (c) free ground area (FGA), (d) areas of the exterior and interior walls (AEW and AIW). Additionally, the required overall resources are calculated with the help of the materials needed by each of the trades. For this, ratios are taken from previously published technical literature (e.g. Spranz 2003). Further input parameters are included: (e) proportion of solids, (f) proportion of formwork, (g) proportion of reinforcements.

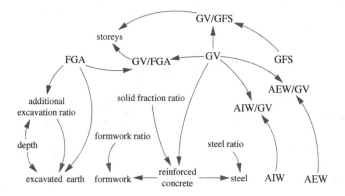

Fig. 3.2 Fundamental project model

To determine the necessary amount of material to be excavated, the following parameters are considered: (h) depth, (i) additional excavation ratio. The parameter additional excavation ratio should give room for additional excavation work, which may be necessary depending on the depth and the resulting need to form berms. In the end, the output parameters of this model are quantities given for (a) excavated earth, (b) formwork, (c) reinforced concrete, (d) steel, (e) brickwork (interior and exterior). Further parameters are relative auxiliary variables which may be useful in the later steps of the modeling process.

3.4   Identification of the Relevant System Parameters

The model *construction project* will be additionally specified in the due course of development under inclusion of the SDL PO (p. 26) and in line with the processes and upcoming tasks. In due consideration of the usual and conventional planning of preparations and operating procedures in the construction industry, a hypothetical and rough draft of a generic construction project will be introduced in the following. In this, the first project definitions of the underlying model are filled with more details. To prepare the following modeling of the units, indispensable information like, e.g., processes, resources and characteristics, defined by the SDL notation, is located with the help of the SDL PO (Fig. 3.3).

At the beginning of the actual production, the setup of the construction site requires attention. This part of the modeling is relevant because the construction site setup can have significant influence on the general expenses of the construction in the area finance and governance (FG). The choice of the individual elements of the construction site setup is based on Schach and Otto (2008) and is displayed in Table 3.1.

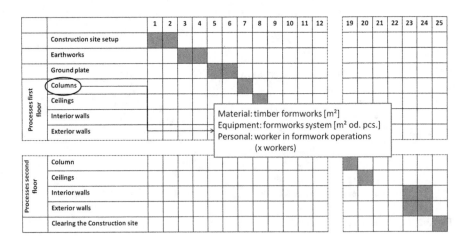

Fig. 3.3 Rough operation chart of project model

Table 3.1 Resources of construction site setup

Process	Material	Equipment	Personnel
Construction site setup	–	Crane	Crane operator
Construction site setup	–	Personnel container	Workers
Construction site setup	–	Foreman container	Foreman
Construction site setup	–	Construction management office	Construction manager
Construction site setup	–	Sanitary facilities	All
Construction site setup	–	First aid ward and equipment	All
Construction site setup	–	Office and meeting rooms	Various
Construction site setup	–	Equipment storeroom	All
Construction site setup	Various	Construction material storage	All
Construction site setup	Various	Operating supplies	All
Construction site setup	–	Construction site workshop	Various
Construction site setup	–	Laboratories	Various

Table 3.2 Resources earthworks

Process	Material	Equipment	Personnel
Excavation	Ground	Excavator	Operator excavator
Transport	Ground	Truck	Truck driver

After the setup of the construction site, the earthworks begin. These can usually be divided in the subsections of removal (excavation), hauling (transport) and disposal (Bauer 2007), whereas the first two of these will be in the main focus of interest. The excavated earth is expected to be unloaded and disposed of without any further costs at the end of the transport. The equipment and the personnel is assigned accordingly (Table 3.2).

The monitored construction of the building shell is segmented in the two phases concrete/reinforced concrete construction (ground plate and processes ground floor to second floor) and brickwork (only processes ground floor to second floor). The included process groups of concrete/reinforced concrete construction are formwork, reinforcing and placing of concrete (Bauer 2007). In this process a distinction is necessary between timber formwork (used, among other areas, in bridge construction) and system formwork (used, among other areas, in general surface construction). In addition, the storage of resources is already allowed in this phase of development (Table 3.3).

Based on the relevant system parameters which were identified in this phase, the key variables in the sense of the AMCA are elaborated in the following.

3.5 Atoms

In general, each single identified resource presents an atom with multiple attributes and can be formally described according to the SDL notation (for details refer to p. 27) as shown in the following.

Table 3.3 Resources building shell

Process	Material	Equipment	Personnel
Formworks	Timber	Formworks system	Worker in formwork operations
Reinforcing	Steel	–	Worker in reinforcement operations
Concrete placement	Concrete	–	Concrete worker
Concrete pumping	Concrete	Concrete pump	Operator of concrete pump
Brickwork	Bricks/mortar	–	Bricklayer
Storage	Various	Various	Warehouseman

3.5.1 Materials

The atoms of the first three materials are listed exemplarily in the following; the remaining materials are described in a similar way.

Excavated earth:

$$a_{MA,1} := a\left(CPD, p_{excavated\,earth}, F(p)\right) \in f(p)\,[u(p)] \tag{3.1}$$

$p_{excavated\,earth}(spec.\,weight\,solid, spec.\,weight\,loose, spec.\,costs, amount, \ldots)$;
$F(p) \in [FG, OP]$; $f(p) \in [stock, const, var]$;
$u(p) \in \left[m^3, to/m^3, to, Euro/m^3, m^3/t\ldots\right]$, $t = $ time unit

(Timber) formworks:

$$a_{MA,2} := a\left(CPD, p_{(timber)\,formworks}, F(p)\right) \in f(p)\,[u(p)] \tag{3.2}$$

$p_{Formworks}(weight, spec.\,costs, amount, \ldots)$;
$F(p) \in [FG, OP]$, $f(p) \in [stock, const, var]$;
$u(p) \in \left[m^2, to, Euro/t, Euro/m^2 \ldots\right]$, $t = $ time unit

Steel:

$$a_{MA,3} := a\left(CPD, p_{steel}, F(p)\right) \in f(p)\,[u(p)] \tag{3.3}$$

$p_{steel}(weigjt, spec.\,costs, amount, \ldots)$;
$F(p) \in [FG, OP]$; $f(p) \in [stock, const, var]$;
$u(p) \in [to, Euro/to, to/t \ldots]$, $t = $ time unit

Depending on the material, atoms are available for the illustration of storage capacities as shown in the following.

3.5.2 Storage

Storage:

$$a_{ST,x} := a\left(CPD, p_{storage}, F(p)\right) \in f(p)\,[u(p)] \tag{3.4}$$

x ∈ [*material code*]

MA: 1 = soil, 2 = formwork, 3 = steel, 5 = brickwork (not storable 4 = concrete);

$p_{storage}$ (*volume, area, material, costs,...*);

F(p) ∈ [*FG, OP*]; f(p) ∈ [*stock, const, var*];

$u(p) \in \left[m^3, m^2, Euro/m^3, Euro/m^2 \ldots\right]$

Apart from the storage capacities, according equipment resources have to be defined together with the materials.

3.5.3 Equipment

The atoms of the first three types of equipment are listed as examples. The rest of the equipment is described accordingly.

Excavator:

$$a_{DE,1} := a\,(CPD, p_{excavator}, F(p)) \in f(p)\,[u(p)] \tag{3.5}$$

$p_{excavator}$ (*depreciation, interest, repair, fuel consumption, performance value, number,...*);

F(p) ∈ [*SO, FG, RD, OP*]; f(p) ∈ [*stock, const, var*];

$u(p) \in \left[number, Euro, m^3, l/t, t/m^3, Euro/t, \ldots\right]$, t = time unit

Truck:

$$a_{DE,2} := a\,(CPD, p_{truck}, F(p)) \in f(p)\,[u(p)] \tag{3.6}$$

p_{truck} (*depreciation, interest, repair, fuel consumption, performance value, number,...*);

F(p) ∈ [*SO, FG, RD, OP*]; f(p) ∈ [*stock, const, var*];

$u(p) \in \left[number, Euro, m^3, l/t, t/m^3, Euro/t, \ldots\right]$, t = time unit

Concrete pump:

$$a_{DE,3} := a\left(CPD, p_{concrete\,pump}, F(p)\right) \in f(p)\,[u(p)] \tag{3.7}$$

$p_{concrete\,pump}$ (*depreciation, interest, repair, fuel consumption, performance value, number,...*);

F(p) ∈ [*SO, FG, RD, OP*]; f(p) ∈ [*stock, const, var*];

$u(p) \in \left[number, Euro, l/t, t/m^3, Euro/t, \ldots\right]$, t = time unit

The containers which are necessary for the construction site setup are described in the following. These can serve as examples for the definition of further units.

Personnel container:

$$a_{DE,5} := a\left(CPD, p_{personnel\,container}, F(p)\right) \in f(p)\,[u(p)] \tag{3.8}$$

Foreman container:

$$a_{DE,6} := a\left(CPD, p_{foreman\ container}, F(p)\right) \in f(p)\ [u(p)] \tag{3.9}$$

Construction management office:

$$a_{DE,7} := a\left(CPD, p_{construction\ management\ office}, F(p)\right) \in f(p)\ [u(p)] \tag{3.10}$$

Sanitary facilities:

$$a_{DE,8} := a\left(CPD, p_{sanitary\ facilities}, F(p)\right) \in f(p)\ [u(p)] \tag{3.11}$$

Generally true for the elements of the construction site setup:

p_i (*depreciation, interest, repair, power consumption, manageable employees, number, . . .*);
with $i \in$ [*personnel container, foreman container, construction management office, . . .*] ;
$F(p) \in [SO, FG, RD, OP]$; $f(p) \in [stock, const, var]$;
$u(p) \in [number, Euro/Stk., to, Euro/to, Euro/t \ldots]$, $t =$ time unit

The system formworks must not be forgotten, it has to be defined as equipment as well.
(System-)formworks:

$$a_{DE,16} := a\left(CPD, p_{(system)\ formworks}, F(p)\right) \in f(p)\ [u(p)] \tag{3.12}$$

$p_{(system)\ formworks}$ (*weight, depreciation, interest, repair, spec. costs, amount, . . .*);
$F(p) \in [FG, OP]$, $f(p) \in [stock, const, var]$;
$u(p) \in \left[m^2, to, Euro/t, Euro/m^2 \ldots\right]$, $t =$ time unit

To finalize the fundamental modeling phase, the required personnel resources have to be defined as well.

3.5.4 Personnel

The examples of an earth-moving plant (excavator) and an industrial employee (bricklayer) are used to demonstrate the definition of the currently needed units. The description of further units follows the same pattern.
Operator excavator:

$$a_{WO,1} := a\left(CPD, p_{operator\ excavator}, F(p)\right) \in f(p)\ [u(p)] \tag{3.13}$$

Bricklayer:

$$a_{WO,7} := a\left(CPD, p_{bricklayer}, F(p)\right) \in f(p)\ [u(p)] \tag{3.14}$$

Generally true for the personnel:

p_i(*working time value, experience, motivation, spec. costs, number, ...*)
with   i   ∈   [*operator excavator, truck driver, operator pump, worker in formworks, ...*] ;
$F(p) \in [SO, FG, HL, OP]$; $f(p) \in [stock, const, var]$;
$u(p) \in [number, m^3/t, Euro/t, ...]$, t = time unit

The effect relationships between the individual atoms and their specific attributes are examined in more detail in the following. As a result, the previously defined atoms are combined and form molecules with their own system structure.

3.6 Molecules

Depending on the processes to be inspected, atoms are combined to form molecules. As the potential properties of atoms and, consequently, the resulting molecules are highly diverse, the following considerations are assigned to different functional areas. To set the structure of the molecules, a further discussion of the possible properties of the individual atoms is necessary. This specification is dependent on the purpose for which the molecule is to be developed.

3.6.1 Operations and Procurement (OP)

3.6.1.1 Earthworks

According to Bauer (2007), earthworks can be subdivided in three main operations or procedural steps: removal, hauling, disposal. This means that the first activity is to loosen the soil and move it onto the means of transport. After the earth has been transported to the desired destination, it is unloaded, fitted and compacted. In the following, the previously defined atoms are used to devise two basic molecules, thereafter the latter being introduced in detail.

Loosen:

$$m_{EA,1} := m\left(CPD, OP, a_{MA,1.1}, a_{DE,1.1}, a_{WO,1.1}\right) \in flow\ [m^3] \tag{3.15}$$

Transport:

$$m_{EA,2} := m\left(CPD, OP, a_{MA,1.2}, a_{DE,2.1}, a_{WO,2.1}\right) \in flow\ [m^3] \tag{3.16}$$

For both molecules (Fig. 3.4 at p. 46) applies that a specific amount of material should be excavated and transported away. The process loosen is dependent on the amount of soil to be removed. The process transport can have the amount of earth

a DE, 1.1:performance value excavator   a WO, 1.1:adjustment to the job operator excavator   a DE, 2.1:performance value truck   a WO,2.1:adjustment to the job truck driver

Fig. 3.4 OP models for loosen and transport

to be removed as a basis, but in case the earth has been stored intermediately and is removed as a whole of in parts later on, it appears reasonable to separately assign the amount of earth to be transported away to the process of hauling. Again it is true for both processes, that the soil to be loosened respectively be removed are additional parameters which have to be considered just like the performance values of the equipment employed for these tasks. Therefore, four more atoms need to be defined. Details are added to the previously defined units as shown in the following.

Earth to excavate:

$$a_{MA,1.1} := a\,(CPD,\,amount\,to\,remove,\,OP) \in stock\,[m^3] \tag{3.17}$$

Earth to transport:

$$a_{MA,1.2} := a\,(CPD,\,amount\,to\,transport,\,OP) \in stock\,[m^3] \tag{3.18}$$

Performance value excavator:

$$a_{DE,1.1} := a\,(CPD,\,performance\,value\,excavator,\,OP) \in const\,[m^3/t] \tag{3.19}$$

Performance value truck:

$$a_{DE,2.1} := a\,(CPD,\,performance\,value\,truck,\,OP) \in const\,[m^3/t] \tag{3.20}$$

with t = time unit.

Furthermore, the staff-related units $a_{WO,1}$ and $a_{WO,2}$ need to be specified within the work molecules. In the due course of the earthworks, a time-dependent effect of experience gained by adjustment to the job may be necessary to consider.

Adjustment to the job operator excavator:

$$a_{WO,1.1} := a\,(CPD,\,adjustment\,to\,the\,job\,factor\,excavator,\,OP) \in var\,[-] \tag{3.21}$$

Adjustment to the job truck driver:

$$a_{WO,2.1} := a\,(CPD,\,adjustment\,to\,the\,job\,factor\,truck,\,OP) \in var\,[-] \tag{3.22}$$

Further staff-related properties are regarded in the section human resources and leadership (Sect. 3.6.2 pp. 49).

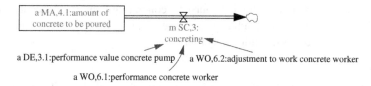

Fig. 3.5 OP model concreting

3.6.1.2   Building Shell

The fundamental units of the sub-process concreting (Fig. 3.5 at p. 47) are illustrated in the following, further units in the sub-processes formworks, reinforcing and brickworks can be developed in the same way. To begin, again a basic molecule is devised and then described in detail. The index SC,3 stands for the third process of the building shell construction (1 = formworks, 2 = reinforcing, 3 = concreting).
Concreting:

$$m_{SC,3} := m \left(CPD, OP, a_{MA,4.1}, a_{DE,3.1}, a_{WO,3.1}, , a_{WO,6.1} \right) \in flow \ [m^3] \qquad (3.23)$$

Similar to earthworks, one type of material and one piece of equipment are considered here ($a_{MA,4}$ and $a_{DE,3}$). Hence, the properties have the following details:
Amount of concrete to be poured:

$$a_{MA,4.1} := a \left(CPD, amount\ of\ concrete, OP \right) \in stock \ [m^3] \qquad (3.24)$$

Performance value concrete pump:

$$a_{DE,3.1} := a \left(CPD, performance\ value\ concrete\ pump, OP \right) \in const \ [m^3/t]$$
$$(3.25)$$

with t = time unit.
A difference, however, is the involvement of two different groups of personnel. The inclusion of a concrete pump requires an operator ($a_{WO,3}$) on the one hand and a concrete worker ($a_{WO,6}$) on the other hand. Regardless of the fact that the concrete pump represents the central and foremost piece of equipment in this process, it appears reasonable to define the performance of the concrete worker as the maximum performance of the pump is primarily limited by the capabilities of the employee using it. The modeler can decide himself or herself (during the design of the simulation) whether this aspect should exert any influence or not.
Performance concrete worker:

$$a_{WO,6.1} := a \left(CPD, performance\ value, OP \right) \in const \ [m^3/t] \qquad (3.26)$$

with t = time unit.

Similar to earthworks, in the due course of concreting a time-dependent process of gaining experience through adjustment to the task may be anticipated. This effect of adjustment to work is considered as follows:

Adjustment to work concrete worker:

$$a_{WO,6.2} := a\,(CPD, adjustment\ to\ work\ factor\ concrete\ worker, OP) \in var\ [-]$$

$$(3.27)$$

Being a central support unit in the process of building shell construction, the crane is modeled as well (Fig. 3.6). Two approaches are given as examples for the determination of the required number of cranes: Determination of crane quantity due to (a) the ratio of employees per crane and (b) performance of building shell construction. Taking into account the number of employees per crane follows the assumption that one crane (depending on the type of construction) can only support a limited number of workers on the construction site (Hoffmann 2006). Thus, one more atom needs to be defined for the modeling of the crane.

Workers per crane:

$$a_{DE,4.1} := a\,(CPD, workers\ per\ crane, OP) \in const\ [workers/crane] \qquad (3.28)$$

The ratio of workers per crane results in the number of cranes I:

$$m_{DE,4.6} := m\,(CPD, OP, m_{WO,0}, a_{DE,4.1}) \in const\ [qty.] \qquad (3.29)$$

with the sum of workers:

$$m_{WO,0} = \sum workers = m(CPD, OP, a_{WO,4.5}, a_{WO,5.5}, \qquad (3.30)$$

$$a_{WO,6.5}, a_{WO,7.5}, a_{WO,10.5}, a_{WO,11.5}) \in const\ [qty.]$$

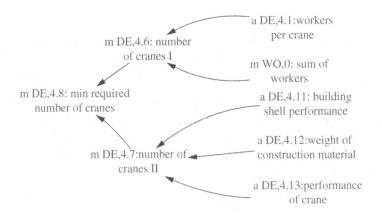

Fig. 3.6 OP model number of cranes

It is the decision of the model designer whether the sum of the employees considers a supervisor respectively a crane operator or annual leave and the number of staff reported sick. As the basic project model includes the gross volume (for details refer to p. 39), the number of cranes can alternatively be determined with the help of the building shell performance (Hoffmann 2006).

The building shell performance results in the number of cranes II:

$$m_{DE,4.7} := m\left(CPD, OP, a_{DE,4.11}, a_{DE,4.12}, a_{DE,4.13}\right) \in const \; [qty.] \qquad (3.31)$$

with building shell performance:

$$a_{DE,4.11} := a\left(CPD, gross\, volume, OP\right) \in const \; [m^3/t] \qquad (3.32)$$

Weight of construction material:

$$a_{DE,4.12} := a\left(CPD, construction\, material\, weight, OP\right) \in const \; [kN/m^3] \qquad (3.33)$$

Performance of crane:

$$a_{DE,4.13} := a\left(CPD, crane\, performance, OP\right) \in const \; [kN/t] \qquad (3.34)$$

with t = time unit.

Therefore, the following value is set for the calculation of the required number of cranes:

$$m_{DE,4.8} := max\begin{pmatrix} m_{DE,4.6} \\ m_{DE,4.7} \end{pmatrix} \in const \; [qty.] \qquad (3.35)$$

3.6.2 Human Resources and Leadership (HL)

3.6.2.1 Earthworks

The personnel-related aspects in regard to the earthwork operations need to be elaborated in more detail in the previously defined units. For this aim, the following atoms are configured as examples.

Experience of excavator operator:

$$a_{WO,1.2} := a\left(CPD, experience\, factor\, excavator, HL\right) \in const, var \; [-] \qquad (3.36)$$

Motivation of excavator operator:

$$a_{WO,1.3} := a\left(CPD, motivation\, factor\, excavator, HL\right) \in const, var \; [-] \qquad (3.37)$$

3.6.2.2 Building Shell

For the consideration of personnel-related aspects in the building shell process, the following units may be of interest, too (analog for the other trades):
Experience concrete worker:

$$a_{WO,6.3} := a\,(CPD, experience\,factor\,concrete\,worker, HL) \in const, var\,[-]$$

(3.38)

Motivation concrete worker:

$$a_{WO,6.4} := a\,(CPD, motivation\,factor\,concrete\,worker, HL) \in const, var\,[-]$$

(3.39)

Additionally, the number of employees per trade is especially important for the areas operations and procurement (OP) and finance and government (FG) (analog for the other groups of persons):
Number of concrete workers:

$$a_{WO,6.5} := a\,(CPD, number\,concrete\,workers, HL) \in const, var\,[qty.] \qquad (3.40)$$

3.6.3 Finances and Governance (FG)

The standard price contract is in the center of interest in the currently inspected operative area of the functional level finance and governance. This includes the separate costs of the following items of work: (1) labor costs, (2) costs of construction material and (3) equipment costs (Leimböck and Klaus 2007).

3.6.3.1 Earthworks

Independently of the required performance, the accumulating costs are of paramount interest. To be able to calculate their values, further units are generated on the basis of the previously defined atoms and molecules. The properties of the deployed equipment and machinery have to be included, as the following exemplary development of further units for one piece of equipment demonstrates. Further equipment can be developed in the same pattern. Equipment costs are normally subdivided into the costs of inventory and maintenance on the one hand (contingency costs) and operating costs on the other hand (Girmscheid and Motzko 2007). Therefore, the molecule $m_{DE,1}$ (with the given example of an excavator) consists of the two molecules $m_{DE,1.1}$ and $m_{DE,1.4}$ which have to be specified.
Equipment costs excavator:

$$m_{DE,1} := m\,(CPD, FG, m_{DE,1.1}, m_{DE,1.4}) \in const, var\,[Euro/t] \qquad (3.41)$$

The contingency costs $m_{DE,1.1}$ consider elements of depreciation and interest rates as well as the repair costs (Leimböck and Klaus 2007) in the molecule $m_{DE,1.3}$. Furthermore, contingency costs include the reinstatement value of the equipment $a_{DE,1.2}$ and also the rates of depreciation and interest $a_{DE,1.3}$.

Contingency costs excavator:

$$m_{DE,1.1} := m\left(CPD, FG, m_{DE,1.2}, m_{DE,1.3}\right) \in const \ [Euro/t] \qquad (3.42)$$

The molecule $m_{DE,1.2}$ contains the calculation of the value for depreciation and interest in dependence on the mean reinstatement value:

Reinstatement value excavator:

$$a_{DE,1.2} := a\left(CPD, reinstatement\ excavator, FG\right) \in const \ [Euro] \qquad (3.43)$$

Depreciation and interest rate excavator:

$$a_{DE,1.3} := a\left(CPD, depriciation\ and\ interest\ rate\ excavator, FG\right) \in const \ [\%/t] \qquad (3.44)$$

Depreciation and interest value excavator:

$$m_{DE,1.2} := m\left(CPD, FG, a_{DE,1.2}, a_{DE,1.3}\right) \in const \ [Euro/t] \qquad (3.45)$$

with t = time unit;

Similarly, maintenance costs are calculated as percentage of the mean reinstatement value in the molecule $m_{DE,1.3}$:

Repair rate excavator:

$$a_{DE,1.4} := a\left(CPD, repair\ rate\ excavator, FG\right) \in const \ [\%/t] \qquad (3.46)$$

Repair value excavator:

$$m_{DE,1.3} := m\left(CPD, FG, a_{DE,1.2}, a_{DE,1.4}\right) \in const \ [Euro/t] \qquad (3.47)$$

with t = time unit;

It has to be considered that the contingency costs are usually determined for a period of one month. Hence, the introduction of an auxiliary variable is necessary (factor 1 for simulation step), which adapts the result of molecule $m_{DE,1.1}$ according to the time increments used in the simulation (days, weeks, months, quarters).

$$a_{AUX,1} := a\left(CPD, factor\ 1\ for\ simulation\ step, FG\right) \in const \ [Mo/t] \qquad (3.48)$$

To specify the operational costs in relation to time, atoms are devised from the following parameters: (a) engine power, (b) fuel consumption, (c) fuel costs and (d) lubricant surcharge.

Engine power excavator:

$$a_{DE,1.5} := a\,(CPD, engine\,power\,excavator, FG) \in const\,[Kw] \qquad (3.49)$$

Fuel consumption excavator:

$$a_{DE,1.6} := a\,(CPD, spec.\,fuel\,consumption\,excavator, FG) \in const\,[l/Kwh]$$
$$(3.50)$$

Fuel costs excavator:

$$a_{DE,1.7} := a\,(CPD, fuel\,costs\,excavator, FG) \in const\,[Euro/l] \qquad (3.51)$$

Lubricant surcharge excavator:

$$a_{DE,1.8} := a\,(CPD, lubricant\,surcharge\,excavator, FG) \in const\,[\%] \qquad (3.52)$$

Finally, the costs of the operating fluids are summarized in the following molecule.

Costs of operating fluids excavator:

$$m_{DE,1.4} := m\,(CPD, FG, a_{DE,1.5}, a_{DE,1.6}, a_{DE,1.7}, a_{DE,1.8}) \in const\,[Euro/h]$$
$$(3.53)$$

Similarly to the contingency costs of the equipment, the costs of the operating fluids are usually calculated per operating hour. Therefore, it is necessary again to introduce a further auxiliary variable ($a_{AUX,2}$), which adapts the results of the molecule $m_{DE,1.4}$ according to the time increments used in the simulation (days, weeks, months, quarters).

$$a_{AUX,2} := a\,(CPD, factor\,2\,for\,simulation\,step, FG) \in const\,[h/t] \qquad (3.54)$$

In case the costs of the excavator operator are supposed to be allocated to the equipment costs, the already introduced atom $a_{WO,1}$ has to be specified with the individual properties of the operator.

Costs of operator excavator:

$$a_{WO,1.1} := a\,(CPD, , spec.\,costs\,operator\,excavator, FG) \in const\,[Euro/t]$$
$$(3.55)$$

with t = time unit.

Thus, the following molecule describes the overall costs of the equipment:

Costs excavator:

$$m_{EA,1.1} := m\,(CPD, FG, m_{DE,1}, a_{WO,1.1}) \in const\,[Euro/t] \qquad (3.56)$$

with t = time unit

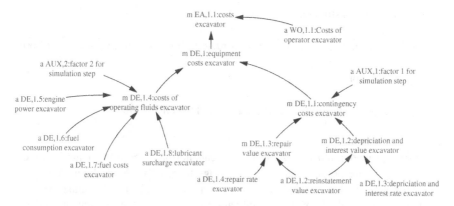

Fig. 3.7 FG model of one item of construction equipment (example excavator)

An overview of the units developed in this context and their relationships is shown in Fig. 3.7.

Based on the previous units, further construction equipment can be developed accordingly. The required data can be acquired from equipment data sheets or literature (in Germany e.g. Baugeräteliste 2007).

Apart from the equipment costs, it has to be taken into account that the removed material causes costs itself. In case the excavated earth is to be disposed, the following applies:

Spec. costs earth disposal:

$$a_{MA,1.3} := a\left(CPD, spec.\ costs\ earth\ disposal, FG\right) \in const\ \left[Euro/m^3\right] \quad (3.57)$$

Costs earth disposal:

$$m_{EA,1.2} := m\left(CPD, FG, a_{MA,1.1}, a_{MA,1.3}\right) \in const\ \left[Euro/t\right] \quad (3.58)$$

with t = time unit.

In case the excavated material is to be stored temporarily, storage costs need to be included in the calculation. As storage capacities are not necessarily equal to the amount of the excavated material ($a_{MA,1.1}$) (they may be larger or smaller), it is required to define the capacity of a storage on basis of the atom $a_{ST,1}$ in more detail.

Capacity earth storage:

$$a_{ST,1.1} := a\left(CPD, capacity\ earth\ storage, OP\right) \in const\ \left[m^3\right] \quad (3.59)$$

Analog to the molecule $m_{EA,1.2}$, the costs for the earth storage are calculated in the following. However, the storage capacity is considered and not the excavated amount of earth.

Spec. costs earth storage:

$$a_{MA,1.4} := a\,(CPD, spec.\ costs\ earth\ storage, FG) \in const\ [Euro/m^3] \qquad (3.60)$$

Costs earth storage:

$$m_{EA,1.3} := m\,(CPD, FG, a_{MA,1.4}, a_{ST,1.2}) \in const\ [Euro/t] \qquad (3.61)$$

with t = time unit.

As the ground removed in earthworks can normally be used again in other projects, there are possibly additional earnings to be considered. As it cannot be taken for granted that the material can be reused and sold for the same price at the point of time of removal, a specification of the re-usable amount of earth is necessary, similar to the earth storage. A further atom is specified on the basis of the previously defined atom $a_{MA,1}$.

Reusable amount of earth:

$$a_{MA,1.5} := a\,(CPD, amount\ reusable\ earth, OP) \in stock\ [m^3] \qquad (3.62)$$

Based on this, the specific earnings can be included to calculate the overall earnings in the area of reutilization.

Spec. earnings earth:

$$a_{MA,1.6} := a\,(CPD, spec.\ earnings\ earth, FG) \in const\ [Euro/m^3] \qquad (3.63)$$

Earnings earth:

$$m_{EA,1.4} := m\,(CPD, FG, a_{MA,1.5}, a_{MA,1.6}) \in const\ [Euro/t] \qquad (3.64).$$

with t = time unit.

The above-mentioned units allow a diversity of combinations to illustrate disposal, storage of earth or its reutilization. A graphic overview of the units in this context is shown in Fig. 3.8.

A combination of the previously developed molecules can be used to define single components as to be seen in Sect. 3.7 (p. 57 ff); these are the first mini-systems.

Fig. 3.8 FG model for disposal, storage and reutilization of earth

3.6.3.2 Building Shell

Like in the area of operations and procurement (p. 45), the main interest is the costs of the concreting process again. The units developed in this context can be modeled for the other trades following the same procedure. Taking into account the previously devised units for the calculation of the equipment costs in the area of earthworks, the equipment costs are determined for the concrete pump involved in the concreting process. Opposed to that, the costs of material and personnel need to be differentiated much more than before. To specify the expenses for the material, the price of concrete and the quantity of it to be deployed are taken into account.

Price of concrete:

$$a_{MA,4.2} := a\left(CPD, spec.\,costs\,concrete, FG\right) \in const\,[Euro/m^3] \qquad (3.65)$$

Concrete costs:

$$m_{MA,4.1} := m\left(CPD, FG, a_{MA,4.1}, a_{MA,4.2}\right) \in const\,[Euro/t] \qquad (3.66)$$

with t = time unit.

A possible and process-related concrete loss has to be considered and is included in an atom with a percentage factor of the overall quantity of concrete to be processed.

Concrete loss:

$$a_{MA,4.3} := a\left(CPD, factor\,of\,concrete\,loss, FG\right) \in const\,[\%] \qquad (3.67)$$

The atom $a_{MA,4.3}$ causes an increase of the needed quantity of concrete ($a_{MA,4.1}$), resulting in an increase of the construction material costs. Apart from the material costs, the personnel costs represent another part of the overall expenses. Therefore, the labor costs for this process are modeled in relation to the amount of concrete to be deployed $a_{MA,4.1}$, the performance value of the personnel $a_{WO,6.1}$, the number of the assigned employees $a_{WO,6.5}$ and their pay $a_{WO,6.3}$ (Fig. 3.9).

Labor costs concreting:

$$m_{WO,6.1} := m\left(CPD, FG, a_{MA,4.1}, a_{WO,6.1}, a_{WO,6.3}\right) \in const\,[Euro] \qquad (3.68)$$

Wages concreting:

$$a_{WO,6.3} := a\left(CPD, spec.\,wages\,concrete, FG\right) \in const\,[Euro/t] \qquad (3.69)$$

with t = time unit.

More basic values are needed to describe the construction site setup as part of the building shell process. Based on the previously defined atom $a_{DE,5}$ to $a_{DE,15}$ (refer to p. 44), the following additional units are necessary:

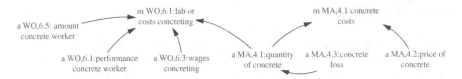

Fig. 3.9 FG model concreting

Workers per container:

$$a_{DE,5.1} := a\,(CPD, workers/container, OP) \in const\ [workers/qty.] \qquad (3.70)$$

Foremen per container:

$$a_{DE,6.1} := a\,(CPD, foremen/container, OP) \in const\ [foreman/qty.] \qquad (3.71)$$

Construction manager per container:

$$a_{DE,7.1} = a(CPD, construction, manager/container, OP) \qquad (3.72)$$
$$\in const\ [construction\ manager/qty.]$$

Persons per sanitary facility:

$$a_{DE,8.1} := a\,(CPD, Persons/container, OP) \in const\ [persons/qty.] \qquad (3.73)$$

Persons per first aid ward:

$$a_{DE,9.1} := a\,(CPD, persons/first\ aid\ ward, OP) \in const\ [persons/qty.] \qquad (3.74)$$

Including the previously defined number of persons per trade ($a_{WO,4.5}$, $a_{WO,5.5}$, $a_{WO,6.5}$ and $a_{WO,7.5}$) respectively the supervisory personnel ($a_{WO,10.5}$ and $a_{WO,11.5}$), the number of required elements for the construction site setup can be specified with the following molecules:

Number of worker containers:

$$m_{DE,5.1} := m\,(CPD, FG, a_{DE,5.1}, a_{WO,4.5}, a_{WO,5.5}, a_{WO,6.5}, a_{WO,7.5}) \in const\ [qty.]$$
$$(3.75)$$

Number of foreman containers:

$$m_{DE,6.1} := m\,(CPD, FG, a_{DE,6.1}, a_{WO,10.5}) \in const\ [qty.] \qquad (3.76)$$

Number of construction manager containers::

$$m_{DE,7.1} := m\,(CPD, FG, a_{DE,7.1}, a_{WO,11.5}) \in const\ [qty.] \qquad (3.77)$$

Number of containers/sanitary facilities:

$$m_{DE,8.1} := m(CPD, FG, a_{DE,8.1}, a_{WO,4.5}, a_{WO,5.5}, a_{WO,6.5}, \quad (3.78)$$

$$a_{WO,7.5}, a_{WO,10.5}, a_{WO,11.5}) \in const \ [qty.]$$

Number of required containers with first aid ward and equipment likewise ($m_{DE,9.1}$).

According to Poloczek (2013), the majority of construction businesses in the surface construction industry rents containers for social purposes, offices and storage facilities. Therefore, the specific and time-related costs per unit ($a_{DE,5.6}$) can be used to calculate the expenses for the above-mentioned elements as shown in the following example (Further elements of a construction site setup to be modeled accordingly.):

Spec. costs personnel containers:

$$a_{DE,5.6} := a\,(CPD, spec.\,costs\,personnel\,container, FG) \in const\,[Euro/t\,per\,qty.]$$
$$(3.79)$$

Costs personnel container:

$$m_{DE,5.2} := m\,(CPD, FG, m_{DE,5.1}, a_{DE,5.6}) \in const\,[Euro/t] \quad (3.80)$$

with t = time unit.

Alternatively, the costs for the construction site containers can be modeled similarly to the previously illustrated, more general approach for construction equipment. The relation between the above-mentioned units is shown by the example of four molecules in Fig. 3.10. However, on condition that there is or are more than one container of one type, this example follows the assumption that all containers of this one type are identical. Nevertheless, the open concept of the SDL approach makes different realizations possible on the basis of existing units, i.e. further specification can also allow different construction site containers per type.

3.7 Components

This phase of the modeling is characterized by the linking of molecules and atoms to form components which have their own internal problem-related processing logic. Building up on the already modeled units, these components depict (a) the construction site setup, (b) the earthworks and finally (c) the process of the building shell construction.

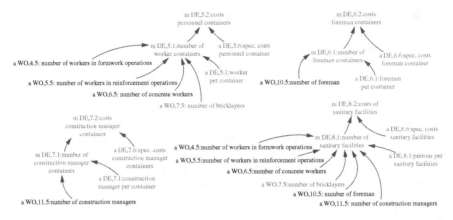

Fig. 3.10 FG models construction containers

3.7.1 Costs of Construction Site Setup

The component *costs of construction site setup* ($c_{SF,1}$) is primarily composed of the two central parts (a) container and (b) crane. Some of these costs are fixed and some are variable. Their mix can affect the simulation outcome and risk modeling.

$$c_{SF,1} := c(CPD, FG, m_{DE,4.5}, m_{DE,5.2}, m_{DE,6.2}, m_{DE,7.2}, \qquad (3.81)$$

$$m_{DE,8.2}, m_{DE,4.5}, m_{DE,4.8}) \in const, var\ [Euro/t]$$

with $t =$ time unit.

The above-mentioned components build up on the previously modeled units and are dependent on the employees of the construction site. Beyond this, the number of required industrial workers is also dependent on the amount of material to be deployed and used (this relationship will be considered in the subsequent generation of the project model, for details refer to p. 75 ff). Thus, the component construction site setup comprises units from the area operations and procurement on the one hand and from finances and leadership on the other hand. The further modeling is carried out on the assumption that all elements of the construction site setup (like containers or cranes) are identical in their categories if there is more than one of them. Based on the atoms defined in Sect. 3.5, the separate elements of the construction site setup are primarily linked to the groups of persons which influence the number of elements. The sum of employees was already given on page 48 with the unit $m_{WO,0}$ and will be included in the subsequent procedure of the modeling. The following units are set as completely independent of the personnel:

- costs meeting rooms ($m_{DE,10.2}$)
- costs tools and hardware storage ($m_{DE,11.2}$)
- costs construction material storage ($m_{DE,12.2}$)

- costs operating supplies storage ($m_{DE,13.2}$)
- costs construction site workshops ($m_{DE,14.2}$)
- costs laboratories ($m_{DE,15.2}$)

The modeler can decide whether and in which number the unit(s) should be represented in the model by choosing the appropriate quantity of units. If, e.g. no construction material storage is planned, the unit $a_{DE,12.1}$ can be set to zero (0) or can be removed entirely. However, if this quantity is dependent on further input parameters (e.g. the amount of building shell construction to be performed), the atom $a_{DE,12.1}$ can be expanded to the molecule $m_{DE,12.1}$, in which the unit quantity is specified by the other parameters.

Further costs for the preparation and realization of a road transport infrastructure or other routes of transport, the construction site safety or the provision and disposal of media are incorporated as unit $a_{DE,16}$ and take the other costs of the construction site setup into account as one flat sum per simulation step. In regard to the modeling of the crane, its operation and standby displays similarities to the already introduced modeling of equipment, but fuel type and quantity is not based on the consumption of diesel, but on electricity. This aspect is going to play a role in Sect. 4.1 on page 75 ff. Furthermore, the auxiliary unit $a_{AUX,3}$ is introduced during the determination of the required number of cranes in the component construction site setup. This auxiliary reflects the fact that not all workers of one construction site can access the support of the crane simultaneously. The unit can be set to values between one (1) and zero (0), whereas the value one means that all workers need to be considered in the calculation of the required crane quantity and, respectively, a value of 0.5 means that only half of the workers need to be considered. Finally, it is the model designer who can decide whether this factor plays a role at all and to which degree.

$$a_{AUX,3} := a\,(CPD, parallelization\,factor, FG) \in const, var\,[-] \qquad (3.82)$$

To illustrate the costs of the construction site setup, the above-mentioned considerations lead to the component $c_{SF,1}$, shown in Fig. 3.11.

3.7.2 Costs of Earthwork

When modeling the costs of earthworks, it has to be determined whether a direct relationship exists between the removal of the earth, i.e. the primary piece of equipment in this context, the excavator or some other device with this purpose, and the process of transport realized with the help of the involved means of transport. If removal and means of transport form a logical chain, the required number of transport vehicles has to be established in dependence of the performance value of the removal device, e.g. the excavator. If, however, the loosened and removed earth is temporarily stored and potentially transported away at a later point of time, the number of needed means of transport cannot directly be derived from

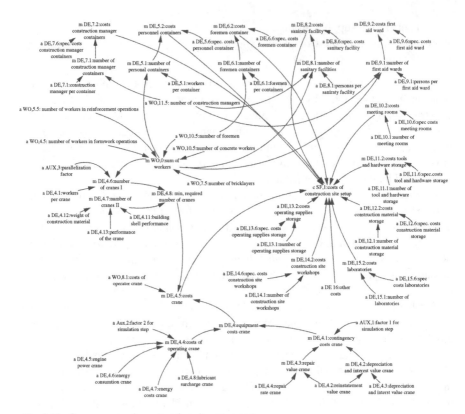

Fig. 3.11 Components of construction site setup costs

the performance of the removal equipment. In this case, the process of transport can usually be seen as independent of the loosening and excavation of the material. Hence, two components are introduced in the section of earthworks: (a) *costs of earthworks without interim earth storage* and (b) *costs of earthworks with interim earth storage*. Both units are based on the assumption that the excavating device can either directly load the material onto the transport vehicle or the storage can be filled directly from the excavator without the necessity to realize additional transport processes. Beyond this it is expected that several devices of one employed type, e.g. excavating devices or means of transport, always entail identical technical specifications.

As a consequence of the simulated time increments (hour, day, week, month etc.), it may be reasonable to consider the required time of the transport device. As the subsequent simulation may have time steps ranging from weeks to months, the separate consideration of the loading time as an independent unit does not appear sensible and is therefore included with the performance value of the removal device. Depending on the aim of the simulation, it may be necessary to devise a model with smaller simulation steps. In such cases, the open architecture of CDL offers the possibility to complement existing units with further atoms and/or complete molecules to depict additional boundary conditions and relationships.

As the atom $a_{MA,1.1}$ represents one level, it is required to fix the demanded amount of excavated earth in the component, thus allowing the reference of this value in the subsequent simulation. Hence, the required amount of earth to be excavated is included as a constant value as follows.

Demanded amount of excavated earth:

$$a_{MA,1.0} := a\,(CPD, amount\,of\,excavated\,earth, OP) \in const\,[m^3] \qquad (3.83)$$

Similarly to the example of brickworks on page 20, an end condition has to be established to indicate the completely achieved excavation amount. For this purpose the following auxiliary unit is used. The auxiliary unit $a_{AUX,4}$ can also play a facilitating role in the subsequent modeling of the functional area finance and governance, as it provides the basis for the process parameters schedule performance index (SPI), time estimate at completion (TEAC) and others (Krause and Arora 2008).

$$a_{AUX,4} := a\,(CPD, excavation\,completed, OP) \in const\,[-] \qquad (3.84)$$

In the case that a part of the earth is to be used again, a differentiation between deposited earth and reusable earth ($a_{MA,1.5}$ on p. 54) has to be specified.

Deposited amount of earth:

$$a_{MA,1.7} := a\,(CPD, deposited\,amount\,of\,earth, OP) \in stock\,[m^3] \qquad (3.85)$$

The distribution of the demanded amount of excavated earth is realized with a relative value, which is depicted with the auxiliary unit $a_{AUX,5}$.

Relative value:

$$a_{AUX,5} := a\,(CPD, proportion\,disposal, OP) \in const\,[-] \qquad (3.86)$$

The value for $a_{MA,1.5}$ results out of $a_{MA,1.0} \cdot (1 - a_{AUX,5})$.

The description of the two different groups of earth is based on the assumption, that the first layer of excavated earth is the reusable part and the earth to be disposed will be found in the later stages of the excavation process. To calculate the costs of the involved equipment in the sub-process loosen, the number of excavation devices planned for this task in the area operations and procurement is required.

$$a_{DE,1.9} := a\,(CPD, number\,excavators, OP) \in const, var\,[qty.] \qquad (3.87)$$

As the necessary number of transport devices depends on the performance of the excavating equipment (Hoffmann 2006), the following molecule is devised to calculate this number in direct relation to the loosening process ($m_{EA,1}$), the performance value of the transport device ($a_{DE,2.1}$), the relative proportion of material to be transported (considered by $a_{AUX,5}$ in $a_{MA,1.5}$ and $a_{MA,1.7}$) and a potential adjustment to the job effect of the equipment operator ($a_{WO,2.1}$).

Number of trucks:

$$m_{DE,2.5} := m\left(CPD, OP, m_{EA,1}, a_{DE,2.1}, a_{AUX,5}, a_{WO,2.1}\right) \in const, var \; [qty.]$$
(3.88)

The linking of these units to determine the equipment costs and the required quantity of devices for this process results in the overall equipment costs. The transport costs are based on the assumption that the employed means of transport for each route of transport or each destination are identical. Due to this, the number of necessary transport vehicles is added in the according molecules. Nevertheless, when using the units of the CDL, the model designer can decide himself or herself whether this simplification should be applied to the model in development. A classification of the equipment costs, e.g. according to the destination of transport, can be realized without any difficulties by repeated integration of the molecule $m_{EA,2.1}$ in combination with the corresponding number of transport devices.

Costs of excavation:

$$m_{EA,1.5} := m\left(CPD, FG, m_{EA,1.1}, a_{DE,1.9}\right) \in stock \; [Euro]$$
(3.89)

Costs of transport:

$$m_{EA,1.6} := m\left(CPD, FG, m_{EA,2.1}, \sum m_{DE,2.5}\right) \in stock \; [Euro]$$
(3.90)

Taking into account the molecules already introduced in Sect. 3.6.3 on page 54, costs of earth disposal ($m_{EA,1.2}$) and additional earnings of earth reutilization ($m_{EA,1.4}$), the general costs for earthworks without interim earth storage are determined with the help of the following component (Fig. 3.12). Note: the unit $a_{MA,1.2}$ should serve (a) better understanding, (b) as confirmation of plausibility ($a_{MA,1.2}$ has to be zero at all times, as nothing is stored) and (c) for the facilitated preparation of the component earthworks including interim storage. The component $c_{EA,1}$ could also be modeled without $a_{MA,1.2}$.

$$c_{EA,1} := c\left(CPD, FG, m_{EA,1.2}, m_{EA,1.4}, m_{EA,1.5}, m_{EA,1.6}\right) \in stock \; [Euro]$$
(3.91)

In contrast to component $c_{EA,1}$, the following component $c_{EA,2}$ integrates an interim storage of the excavated material. As already mentioned in the introduction of this chapter, the equipment used for the earth removal and the transport devices are not directly linked to each other. This means that the excavated material can be transported away at any point of time and the number of required transport devices ought to be viewed independently of the loosening process. For this, the modeling of $c_{EA,2}$ needs to differentiate three fundamental cases.

- case A: the quantity of available material in the storage is insufficient, i.e. the removal performance is not equivalent to the transport capacity of the vehicles and a transport standstill is the consequence.

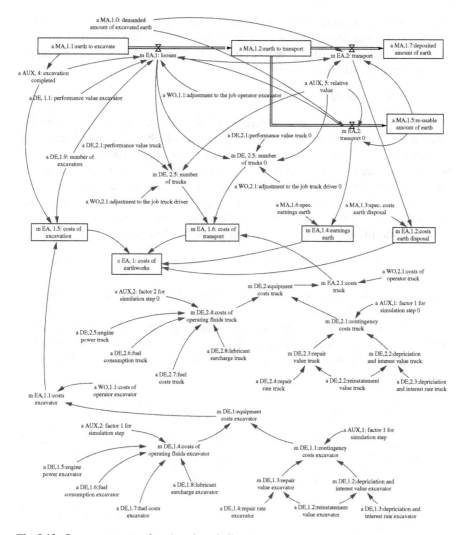

Fig. 3.12 Component costs of earthwork excluding storage

- case B: the removal performance exceeds the transport capacity, i.e. an interim earth storage takes place.
- case C: the removal performance exceeds the transport capacity of the transport equipment, but at the same time it exceeds also the storage capacity, i.e. the removal process is halted.

Analog to the component $c_{EA,1}$, the time which is needed for the loading of the transport device with material from the storage is not explicitly represented in $c_{EA,2}$ (refer to p. 60). In the following, $c_{EA,2}$ is developed on the basis of $c_{EA,1}$. At the beginning, similarly to the required amount of excavation $a_{MA,1.0}$, the maximum capacity of the earth storage needs to be fixed in accordance to the unit $a_{ST,1.1}$ (refer to p. 53) and linked to the unit $m_{EA,1}$. The desired result is the halted excavation in case of a full storage. Furthermore, the previously defined unit $a_{MA,1.2}$ is substituted with the according storage capacity.

$$a_{ST,1.2} := a\,(CPD, capacity\ of\ used\ storage, OP) \in stock\ [m^3] \qquad (3.92)$$

The independence of the processes excavation and transport is achieved by removing the relationships between $m_{EA,1}$ and the number of required transport devices ($a_{DE,2.5}$). With this new independence of the earth removal process, the performance values of the transport devices as well as the potentially applicable effects of adjustment to the job do not have to be considered anymore in the calculation of the required number of devices, but are directly included in the transport processes itself ($m_{EA,2}$). Finally, the costs of the earthworks are expanded with the molecule $m_{EA,1.3}$ to reflect the costs of the earth storage.

$$c_{EA,2} := c\,(CPD, FG, m_{EA,1.2}, m_{EA,1.3}, m_{EA,1.4}, m_{EA,1.5}, m_{EA,1.6}) \in stock\ [Euro]$$
$$(3.93)$$

Figure 3.13 shows the component $c_{EA,2}$ with the changed units only; the units not shown here are analog to $c_{EA,1}$.

The combination of the components $c_{EA,1}$ and $c_{EA,2}$ can aid the development of further components, which could, e.g., illustrate the interim storage of the reusable earth (possibly top soil or surface soil) and the subsequent direct disposal of

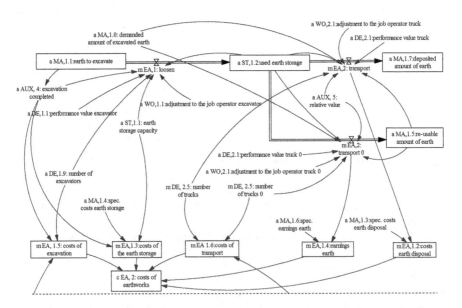

Fig. 3.13 Component costs of earthworks including storage

contaminated earth to a repository. This example shows the potential and flexibility of the CDL, with whose help a multitude of new units can be created on the basis of already existing ones.

3.7.3   Costs Building Shell

Based on the general project model (compare Fig. 3.3 on p. 40) and apart the already discussed processes, the component *costs building shell* ($c_{SC,5}$) incorporates the sub-processes formworks ($c_{SC,1}$), reinforcing ($c_{SC,2}$), concreting ($c_{SC,3}$) and brickworks ($c_{SC,4}$). As mentioned on p. 41, a distinction is necessary in $c_{SC,1}$ between timber formworks ($c_{SC,1.1}$) and system formworks ($c_{SC,1.2}$) which will be considered in the subsequent modeling process. For this aim, an individual component will be modeled for each sub-process, all of which can finally be found in the component costs building shell.

$$c_{SC,5} := c(CPD, FG, c_{SF,1}, c_{SC,2}, c_{SC,3}, c_{SC,4}, \tag{3.94}$$

$$\left\{ \begin{matrix} c_{EA,1} \\ c_{EA,2} \end{matrix} \right\}, \left\{ \begin{matrix} c_{SC,1.1} \\ c_{SC,1.2} \end{matrix} \right\}) \in stock \ [Euro]$$

3.7.3.1   Standard Component

As a preparational step, a standard component is devised to serve as a basis for the modeling of the above-mentioned components. The aim of this development is, on the one hand, the illustration and research of the dependencies and effect-relationships within a production process in the construction context. On the other hand, the standard component should offer practical possibilities, e.g. to guarantee an uninterrupted production with optimized storage expenses through minimal storage capacities and minimized numbers of placed orders. To reach this aim, the component should generate options to influence a reduction or minimization of payments, so that frequencies of orders and the ordered quantities can be varied in such a fashion, that outgoing payments are distributed regularly within a given period of time. A more detailed examination of the cash flow including the debits and credits within a production process is not implemented in this phase. This parameter of productivity shows the flow of financial resources in a set period (usually 12 months) as a result of working capital cycles and is therefore available for investments (Krause and Arora 2008). In the end, the standard component is requested to be able to depict the following:

1. Illustration of a construction-specific production process including performance values, material losses etc.
2. Inclusion of influences exerted by work results of substandard quality and their remedy.

3. Consideration of personnel-related requirements like adjustment to the job, experience, motivation and overtime.
4. Illustration of limited storage capacities including potential material losses due to the storage itself (only applicable if a storage is planned for the according material).
5. Visualization of fundamental supply chains; especially consideration of delays caused by order placement, processing and delivery of material.
6. The storage should be used to capacity and the continuous production process guaranteed by constantly adjusted and monitored economic order quantities.
7. The ordered quantity should be adjustable even in accordance to the predicted production.
8. The point in time of the last order placement should be free to define individually, so that a potentially available remaining quantity of material in the storage can be minimized.
9. A maximum of possibilities should be given to influence and examine the behavior of the component.
10. Illustration of the parameters respectively auxiliaries as, e.g., average storage time, actual production performance or prospective duration of production.
11. Open structure to allow expansion and integration of potentially required additional aspects in the further procedure of the development. This could, for example, include different risks and their influence.

Beyond these demands, the standard component is subject to the following conditions as well:

1. As the maximum period of time for the completion of the construction task is usually limited and set in the contract, the end of the process as set in the model should be seen as fixed.
2. The maximum storage capacity is limited, but can be reduced if necessary.
3. The storage (if existing for the according material) is completely stocked at the beginning of the process.
4. The frequency of order placement should be minimized in the sense of supply chain management (Krüger and Steven 2000).
5. The ordered amount is generally dependent on the fixed end of the process and the order frequency.
6. The stock of inventory, the production performance, potential delays as well as the already present material need to be considered when calculating the order amount.
7. Only one supplier exists for one material, i.e. there is no competition of different suppliers.
8. The supplier can always deliver the requested material at each point in time and in the ordered quantity and quality, independent of possible delays in order processing or delivery itself.
9. The availability of the material on the market is unlimited.
10. The delivered material does not incur quantitative or qualitative losses or damage during the shipping procedure.

11. Quantitative material losses are only caused in the storage or during the handling at work.
12. The material taken from the storage is always of the expected quality. There is no loss of quality during the storage process.
13. Projects with a great amount of change experience lower-than-planned productivity (Ibbs 1997). Result of change are flaws in production and workmanship which will impact experience, motivation or the overtime of the personnel.
14. Substandard performance of the production process itself is, apart from an insufficient speed of production, entirely impossible.
15. Volatile acquisition prices of material neither have an impact on the order amount nor on the order frequency. Therefore, tactical order placement is impossible.
16. Due to the fact, that change is a key contributor to the inconsistent results achieved in projects (Ibbs and Backes 1995), in the simulation the employer can always bring in requests and demand changes of the plan, resulting in necessary adaptations of the implementation and according quantities.

Including the previously defined units, a production molecule serves as a fundamental basis and is successively complemented with further units in accordance to the simulation aim. To fulfill the previously listed requirements of the standard component, it is indispensable to develop further units which consider the aspects supply chain and flaws in workmanship. This results in the design of additional units, especially in the functional area operations and procurement (OP). To allow a clearly laid out overview, the nomenclature of SDL is omitted in Fig. 3.14 in the illustration of the different levels.

The standard component makes it already possible to thoroughly investigate the economic impact of production, storage, delivery etc. The required components for the sub-processes formworks ($c_{SC,1}$), reinforcing ($c_{SC,2}$), concreting ($c_{SC,3}$) and brickwork ($c_{SC,4}$) are developed out of the standard component in the following section.

3.7.3.2 Formworks

A distinction between timber formworks ($c_{SC,1.1}$) and system formworks ($c_{SC,1.2}$) is made in this component. Both need to be present on the construction site in sufficient quantity and at a specific point in time. Thus, in these two components of formworks the same constraints play a role as in the standard component, both in regard to orders and delivery. The component $c_{SC,1.1}$ does not require significant adaptations, as it is completely represented by the standard component. In contrast to this, the use of system formworks has to consider the aspect of reusability. Consequently, a further connection between production and storage has to be established in component $c_{SC,1.2}$, to picture the amount of returned formwork elements and the influence of their duration in the process. To integrate the wear and tear of formwork elements, an additional wear ratio needs to be integrated into

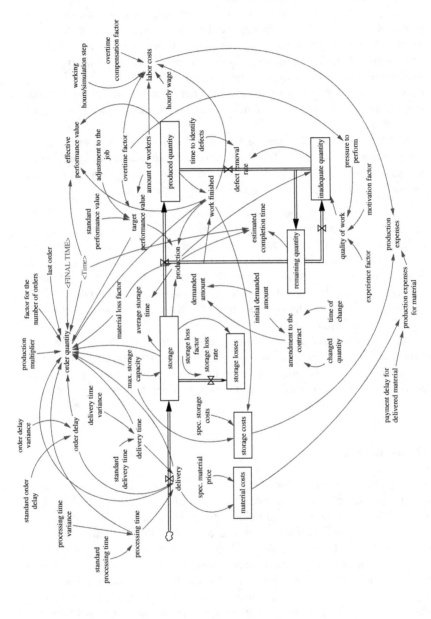

Fig. 3.14 Standard component of a building shell process

the component. Lastly, the material expenses which are only generally characterized as costs of material in the standard component need to be specified in more detail in $c_{SC,1.2}$. Here, two possibilities are available. First, in case the formwork system has been modeled as its own component of equipment costs in the SDL, it can be linked here. Secondly, however, the formwork elements, armaments and shear walls are frequently rented in surface construction (Poloczek 2013), so it is equally possible to include the rental costs per time unit. Figure 3.15 shows the component for the formwork processes with system elements which is derived from the standard component.

As no significant changes are necessary for the component $c_{SC,1.1}$ (timber formwork) in comparison to the standard component, an illustration can be omitted here.

3.7.3.3 Reinforcing

In contrast to the formwork component $c_{SC,1.2}$, the reinforcing process can be derived from the standard component with less adaptations. Even here, the maximal available storage capacity is an important issue for the simulation. Again, the steel has to be available on the construction site in sufficient quantity and quality at the right point in time. Hence, the standard component represents the reinforcing component $c_{SC,2}$ completely and does not need to be adapted. An explicit illustration of this component can therefore be omitted.

3.7.3.4 Concreting

The component concreting ($c_{SC,3}$), however, does not allow storage of the concrete as it is usually delivered to the construction site on time. In general, two possibilities are available if a standard component has already been developed and is in use: (a) irrelevant units are assigned the value zero if the system permits or (b) the irrelevant units are removed completely from the component and the connections are adapted to fit the new system purpose. Even if a standard component is employed, the consistency and reliability of the derived components has to be tested and confirmed in both cases. Figure 3.16 shows the second alternative, in which irrelevant units and connections have been removed and other necessary units were integrated. One of the latter is the concreting cycle, which can be used to define at which intervals concreting should be carried out.

3.7.3.5 Brickwork

The component $c_{SC,4}$ displays similarities to the component $c_{SC,2}$. As a consequence, the standard component reflects this component entirely as well. Due to the reasons which apply to the components $c_{SC,1.1}$ and $c_{SC,2}$, an explicit illustration of this component is omitted here.

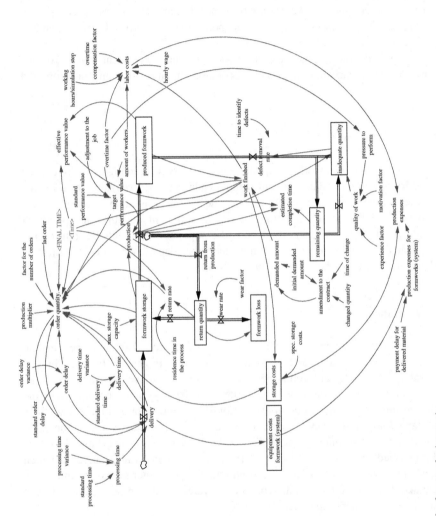

Fig. 3.15 Component for formworks (system)

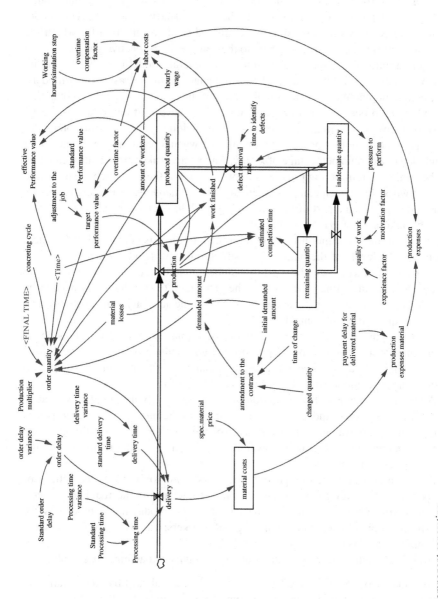

Fig. 3.16 Component concreting

3.8   Project Model

A qualitative project model concept, based on the previously developed units of the CDL, is introduced in the following (Fig. 3.17). This will prepare the ground for the following project simulations in Sect. 4.1, starting page 75. The model depicts a three-story office building in reinforced steel construction for which the building shell is to be constructed. Due to the complexity respectively the multitude of required units, the graphic illustration of the overall model is realized with the USDML (for details refer to Fig. 2.11, starting page 30) In contrast to other models, for example in pipeline construction, this type of illustration takes into account that the individual processes do not necessarily need to have start-end-relations (in the sense of network planning). This means that individual processes can potentially run in parallel or another process can commence when a previous process has been concluded to a certain percentage of x only. An example taken from reality could be the beginning of reinforcing processes in one section of the construction site where the formworks have been completed already, or, alternatively, concreting processes in a section in which a part of the reinforcing work has been done already. For this aim, the auxiliary unit $a_{AUX,6}$ is integrated.

Following the assumption that the represented object is composed of identical floors, i.e. the levels of the building are structurally the same, a query containing the auxiliary unit $a_{AUX,7}$ can be used to include a loop condition in the model. Due to this loop, the modeling processes for the generation of one floor only need to be modeled once and can be repeated with the help of the auxiliary unit $a_{AUX,7}$ as many times as requested in the context.

Both the purpose and the extent of the performance of the project model is determined by the general model purpose already defined on page 37, combined with the already realized demands and additional requirements from the definition of the standard component building shell (refer to p. 65).

Additionally, the model *construction project* respectively the simulation to arise out of it is supposed to independently determine (a) the number of required pieces of equipment (excavators and trucks) and (b) the number of required employees. The user merely has to define the available cycle times for the realization of the necessary tasks and processes.

As a consequence, the component earthworks $c_{EA,1}$ only needs the setting of the period of time for the given task in order to determine the required devices. The following parameters need to be added to the production components for the specification of the required number of workers: (a) week/work cycle and (b) working days/week.

Additionally, the following additional requirements and assumptions apply:

- As a matter of principle, the basic building data are always based on statistical data (e.g. Baukosteninformationszentrum 2008, section office buildings).
- The amount of materials calculated in the original project definition is evenly distributed on the individual floors of the building.

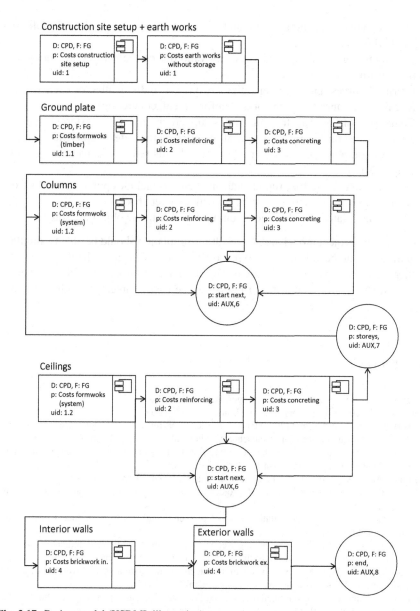

Fig. 3.17 Project model (USDML illustration)

- If quantities increase or decrease, an immediate influence on the time duration of execution will be exerted, but not necessarily on the number of workers..
- The component *costs of construction site setup* ($c_{SF,1}$) inevitably has to be connected to the other, personnel-related production molecules available in the model to guarantee that the workers present on the construction site are also considered in the determination of the construction site setup. Further adjustment of $c_{SF,1}$ is not necessary.

- The originally and contracted amount of work to do is used for the calculation of the workforce.
- If learning curves are applied, these are valid during the entire time duration of the construction.
- Learning curves and learning ratios are viewed individually for each process (excavation, transport, formworks, reinforcement, concreting).
- Containers of the construction site setup are rented from external sources. Therefore, no further investigation of the construction site setup elements in the sense of equipment costs is carried out (for details refer to Baugeräteliste 2007, chapter X).
- Delivery of concrete for ground plate and ceilings includes pumping works from external sources. Concrete installation with concrete buckets.
- The ground plate as well as the columns available in the object are square.

In the end, the project model and the resulting simulation should be able to simulate an unlimited number of different types of buildings with up to 10 floors without further need to intervene in the system. The amount of 10 floors should be sufficient for a first simulation model. More floors can be easily integrated with additional links.

References

Bauer H (2007) Baubetrieb, 3rd edn. Springer, Berlin, Heidelberg
Baugeräteliste (2007) Baugeräteliste 2007. Bauverlag, Gütersloh
Baukosteninformationszentrum (2008) BKI Baukosten 2008 Teil 1: Kostenkennwerte für Gebäude. Baukosteninformationszentrum Deutscher Architektenkammern, Stuttgart
Girmscheid G, Motzko C (2007) Kalkulation und Preisbildung in Bauunternehmen; Grundlagen, Methodik und Organisation. Springer, Berlin, Heidelberg
Hoffmann M (2006) Beispiele für die Baubetriebspraxis, 1st edn. Vieweg+Teubner
Ibbs W (1997) Quantitative Impacts of project change: size issues. J Construct Eng Manag 123(3):308–311
Ibbs W, Backes JM (1995) Quantitative eects of project change. Tech. Rep. 43,2, Construction Industry Institute, Austin
Krause HU, Arora D (2008) Controlling-Kennzahlen - Key Performance Indicators, Zweisprachiges Handbuch Deutsch/Englisch, 1st edn. Oldenbourg, München
Krüger R, Steven M (2000) Supply chain management im spannungsfeld von logistik und management. Wirtschaftswissenschaftliches Studium 29:501–507
Leimböck E, Klaus UR (2007) Baukalkulation und Projektcontrolling; unter Berücksichtigung der KLR Bau und der VOB, 11th edn. Vieweg, Wiesbaden
Poloczek A (2013) Wertorientierte Bewertung von Projekten mit Unikat-Charakter. Ph.D. thesis, Institut für Baubetrieb und Baumanagement, Universität Duisburg-Essen
Schach R, Otto J (2008) Baustelleneinrichtung; Grundlagen-Planung-Praxishinweise-Vorschriften und Regeln. B.G Teubner Verlag/GWV Fachverlage GmbH, Wiesbaden, Wiesbaden
Spranz D (2003) Arbeitsvorbereitung im Ingenieurhochbau. Bauwerk Verlag GmbH, Berlin
VDI 3633 (1996) Simulation von Logistik-, Materialuss- und Produktionssystemen - Begriffsdefinitionen (Entwurf). Beuth Verlag, Düsseldorf

Chapter 4
Implementation of the Model Elements

The practical testing of the previously developed models in form of simulations is demonstrated in the following for the operative level, i.e. using the project model introduced in Sect. 3.8 on page 72. Depending on the intended aims, a diversity of concepts can be realized with the help of the already developed CDL units or their subsequently generated models (Chap. 3, p. 37 ff.).

4.1 Structure of the Project Simulation

Previously, the individual units were qualitatively viewed and registered. To prepare a quantitative examination and the subsequent analysis, the hitherto developed project model concept is implemented in Vensim as described in the following chapter. The simulation does not serve the primary aim to offer a complete, comprehensive and practicable method for the generation of cost-relevant data being as precise as possible. The main focus is rather to draft a realistic project model, which allows elucidating the relationships on the one hand and the influences on the other hand—from a qualitative or a relative and quantitative perspective. In addition to that, the simulation serves as a validation of the model and should help to determine realistic parameters.

As Vensim has already been used since the modeling phase, the contained project model needs to be equipped with corresponding mathematical equations as well as realistic input data in the current simulation phase (The project model consists of more than 750 differential equations). The validation phase comprised both checks executed with a spreadsheet program or as manual calculations, aiming to verify the general validity of the model and also the system stability and integrity. The quantitative database and the further requirements are available in the Appendix 7.1, p.95 ff. While designing a simulation using Vensim, different subdomains and

© The Author(s) 2016
C.K. Karl, W. Ibbs, *Developing Modular-Oriented Simulation Models Using System Dynamics Libraries*, SpringerBriefs in Electrical and Computer Engineering,
DOI 10.1007/978-3-319-33169-0_4

partial models of the comprehensive model can be segmented in so-called "views" (this concept is similar to flags in spreadsheet applications). In the following chapter the developed central graphical user interface (GUI) is introduced. Additional views allow the input of project definitions as well as simulation constants. As examples, individual views are displayed (project definition, construction site setup and building shell process). The design of the simulation resulted in the *Construction Project Flight Simulator* (CPFS) of which the GUI is shown in Fig. 4.1 on 77.

The previously developed CDL units are assigned to 24 views in the CPFS, where the different parameters like, e.g., basic settings of the simulation, learning ratios, operating expenses and performance values can be adjusted in detail. Central views of the CPFS are introduced in brief summaries in the following.

4.1.1   Project Definition

An upgrade of the introduced basic project model (refer to p. 39) appeared necessary and was implemented into the CPFS. The resulting project definition is shown in Fig. 4.2 on 78.

Main difference to the basic model is the extended degree of detail in regard to the materials required by the project, especially dependent on the dimensions of the project as a whole. Due to this, further units are to be included in correspondence to the previously described boundary conditions of the project model (refer to p.72). These units can represent specific dimensions of the ground plate, the ceilings or the columns, for example. To achieve a higher flexibility of the CPFS, the input of a prop grid is possible. Based on these additional inputs, all further required data is generated by the model, e.g. volume and areas for the formworks. According to the previously defined boundary conditions, the project model or the according simulation should be able to depict buildings of up to 10 storeys. Thus, the introduction of building floors as a definition is strictly necessary, too.

4.1.2   Construction Site Setup

The component *costs of construction site setup* ($c_{SF,1}$) requires a special adaptation when integrating it into the CPFS. As this component is dependent on units which, e.g., determine the amount of required personnel for the realization of specific processes, "shadow variables" are introduced (displayed with angle brackets in Vensim). These are variables which exist in one view already, but can be re-used in other views again. The resulting advantage is that central variables only need to be defined once. Hence, this approach makes it possible to change a central variable and automatically cause consequences in each view where it was used as well.

Regarding the view construction site setup, the number of employees per process can be specified in other views in more detail and serve as a basis for the calculation

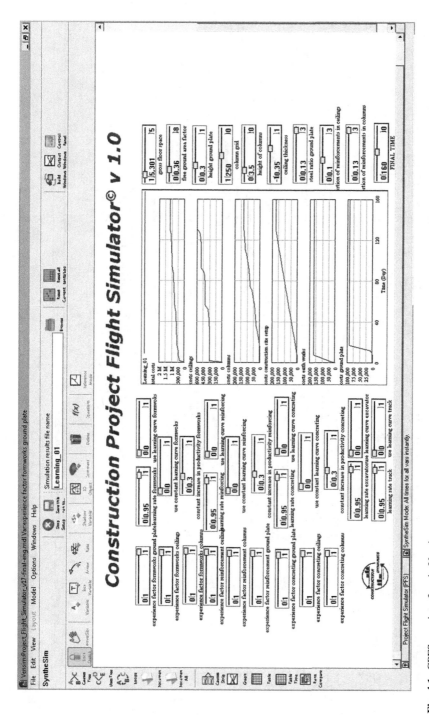

Fig. 4.1 CPFS user interface in Vensim

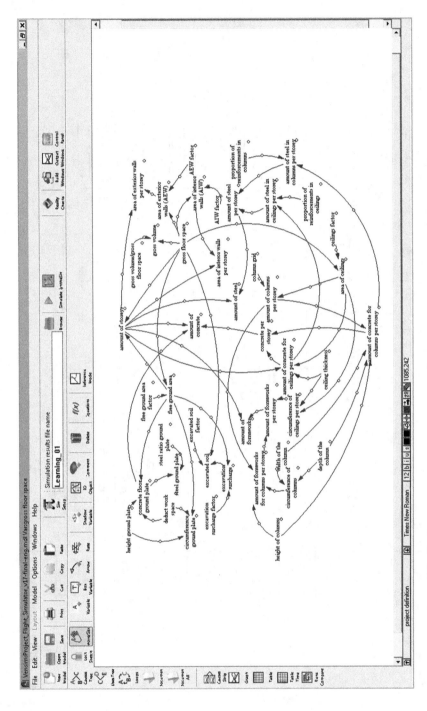

Fig. 4.2 Project definition of the CPFS

of the necessary construction site setup elements (Fig. 4.3). All further variables of the construction site setup can be adapted in this view according to the individual project specifications.

4.1.3   Building Shell Processes

As an example for all building shell processes, the ground plate from the CDL component $c_{SC,1,1}$ (timber formworks) is used to introduce the resulting view (Fig. 4.4). In this view, shadow variables are used to reflect, e.g. the termination of an operation (<excavation finished>), factors influencing the adjustment to the job (<learning curve formworks>), working hours per day (<working hours/day>) as well as the dimensions needed for the calculation of required material (<circumference of formworks for ground plate>). Further units which are only relevant for the shown process are introduced as normal variables, e.g. the parameter of weeks per working cycle. The experiments carried out with the help of CPFS are described in Sect. 5.1 on page 83 ff.

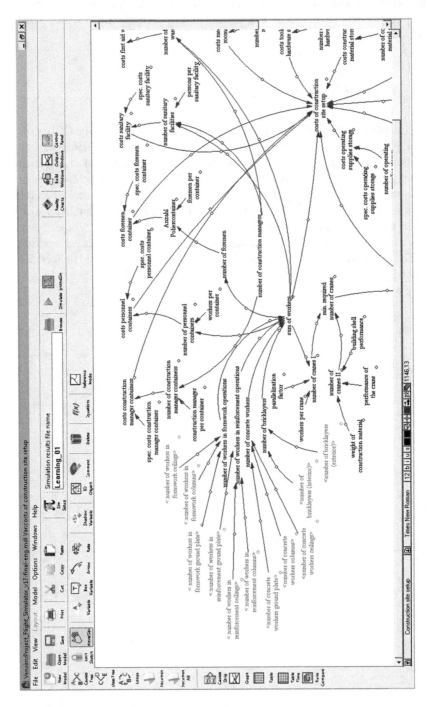

Fig. 4.3 Construction site setup of the CPFS

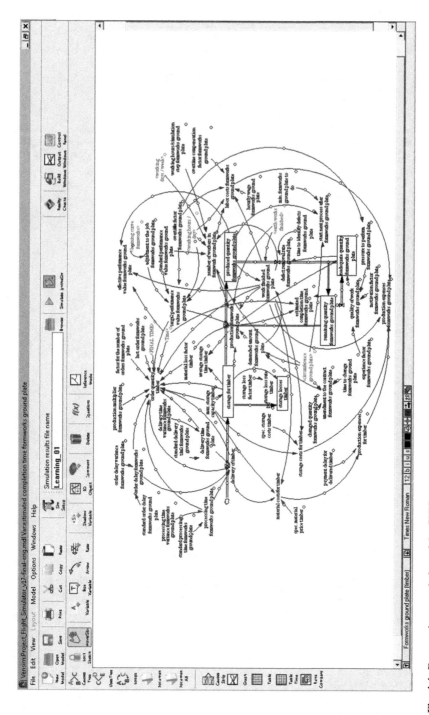

Fig. 4.4 Formworks ground plate of the CPFS

Chapter 5
Selected Results of Prototypical Implementation

Based on the simulation devised in Sect. 4.1, the results of the implemented simulation experiments are introduced and discussed in the following.

5.1 Results of Simulation

The quantitative analyses executed with the following simulation studies serve the verification and validation of the simulation models on the one hand, but, on the other hand, the results should also offer indicators for the examination of specific situations and behaviors of the model and thereby potentially allow the derivation of decision dispositions. Especially the illustration of relationships and dependencies between the different elements should support the evaluation thereof and, even more so, the search for alternatives. Strong emphasis is placed on the overall costs and the duration of the project. A contractually fixed maximum duration for all processes of 160 days must not be exceeded.

An estimated cost function, as shown in Appendix 7.2 on p. 97 ff., is assigned to each of the scenarios for the illustration of the costs development during the project. The functions can provide further assistance in project estimations, research projects or even in teaching-/ learning contexts.

Without any doubt, the devised simulation can be used to create and examine a diversity of scenarios. This chapter introduces 8 selected scenarios. In these scenarios, the system reactions on changes of the input values are explicitly fixed and are executed as deterministic simulations once only. The question settings for the individual deterministic scenarios are the following:

- Scenario S1—Basic: Ideal construction operation, all further scenarios are compared to S1.

© The Author(s) 2016
C.K. Karl, W. Ibbs, *Developing Modular-Oriented Simulation Models Using System Dynamics Libraries*, SpringerBriefs in Electrical and Computer Engineering,
DOI 10.1007/978-3-319-33169-0_5

- Scenario S2—Vocational adjustment: Which influence do familiarization effects have on the work? This scenario could answer the question if it is worth to consider vocational adjustment within the planning and execution of a construction project.
- Scenario S3—Experience: Which influence can be measured in case of a less experienced workforce? This aspect could be important especially for projects with a high amount of subcontractors (like in global projects). In such cases it is necessary to know (a) which performance can be assumed compared to the own work force and (b) is it worth to invest in education for foreign work force and if so, which maximum investment should not be exceeded?
- Scenario S4—Overtime: What is the influence of overtime? This scenario might give hints to decide until which limit mandated overtime can be seen as a common strategy to face delays within construction.
- Scenario S5—Overtime (S4) & Vocational adjustment (S2): In how far does a combination of overtime and adjustment to the work result in the shortest possible period of time needed for the completion? Secondly, in how far can potentially negative effects of overtime be compensated by familiarization effects?
- Scenario S6—Experience (S3) & Variation of time for quality management: Which influence does the quality of the construction management exert, especially in case of less experienced personnel? This scenario takes into account a similar problem like in S3 (e.g. global projects). To face the problem of poor quality as a result of less experienced work force, the management of the project might be a key factor.
- Scenario S7—Overtime (S4) & Variation of time for quality management: Which influence does the quality of the construction management have in case of overtime ordered by it?
- Scenario S8—Crisis management of a bad case scenario: How can a bad case scenario look like, and how does this change when it is influenced by familiarization effects (S2)?

5.1.1   Simulation Experiments

5.1.1.1   Scenario S1: Basis

The basic scenario S1 represents the ideal construction process. All material (except concrete) needed for the first production cycle is present in full quantity at the beginning of the construction. Consequently, no placing of orders is necessary at the start. All employees have maximal experience and motivation, i.e. the work can be done with optimized quality and no revisions of previous processes have to be carried out. Furthermore, losses due to storage handling or wear and tear of the material are excluded. Theoretically possible increases of production due to familiarization to the job are not considered in S1 and no overtime is possible. Both the overall project costs and the overall duration for the regarded processes are basis for the comparison to other scenarios.

5.1.1.2   Scenario S2: Vocational Adjustment

In S2, learning curves are defined for each work process (formworks, reinforcing, concreting) and serve as basis for the examination of effects which the work adjustment can exert on the overall costs and duration. The learning rate is set to 95 % and develops individually for each process from the first activity and then continuously during the whole construction process. The maximal realizable increase of productivity is +30 %. Results: costs −2,79 %, time −8,40 %.

5.1.1.3   Scenario S3: Experience

Scenario S3 focuses the influence of less experienced workers. In this case the work experience of the employees is assumed to be 10 % less than the given standard, resulting in a decrease of the same amount in the execution of assigned tasks. In contrast to S2, the reduction does not exert direct influence on the performance value, but mainly on the quality of the work results. Thus, the lower experience level leads to a faulty workmanship and then to the necessity of revision and corrective activities. Here, the time required for the detection of substandard work results is set to one day, i.e. potentially occurring defective results can be identified and corrected by the construction management on short notice. Result: costs +4,60 %, time +10,00 %.

5.1.1.4   Scenario S4: Overtime

Inspecting the scenario S3, the influence of mandated overtime on the overall project costs and on the overall completion time appears of central interest. To gain insights here, the available working hours per day are increased by 20 %. Assuming that the quality of work remains the same in spite of the mandated overtime, i.e. the overtime does not lead to substandard quality of workmanship, the scenario S4.1 displays the following result: costs −1,39 %, time −10,69 %. If the directive of overtime is linked to the productivity and quality of workmanship (Ibbs and Vaughn 2015), i.e. the overtime has a detrimental effect on the quality of the work and thereby causes the need of revision and repair cycles, the result of scenario S4.2 is this: costs +2,88 %, time −2,29 %. Just as in scenario S3, S4.2 also expects the time required for the detection of substandard work results to be one day.

5.1.1.5   Scenario S5: Overtime (S4) and Vocational Adjustment (S2)

Scenario 5.1 is supposed to represent a good case scenario, combining overtime (S4.1) and vocational adjustment (S2). This scenario, compared to the basic scenario S1, results in the following changes: costs −1,39 % and time −11,45 %. S5.2 poses the question in how far the negative influences of mandated overtime in scenario 4.2 can be balanced with the consideration of vocational adjustment effects. The results of scenario 5.2 in comparison to the basis scenario S1: costs +2,44 %, time −3,05 % and comparing scenario S5.2 to S4.2: costs −0,40 %, time −0,78 %.

5.1.1.6   Scenario S6: Experience (S3) and Variation of Time for Quality Management

Both scenario S3 and S4 relied on the period of one day for the detection of substandard work results. It appears reasonable to elucidate the changes of overall costs and completion time if inexperienced personnel and a less attentive construction management work together. Consequently, scenario S6 examines the influence of the qualification of construction management especially in combination with less experienced construction site personnel. Here, the time required for the detection of substandard workmanship is set to two or three days.

S6.1—time required/detection of substandard quality 2 days: costs +6,36 %, time +12,21 %
S6.2—time required/detection of substandard quality 3 days: costs +9,07 %, time +19,85 %

5.1.1.7   Scenario S7: Overtime (S4) and Variation of Time for Quality Management

Similarly to scenario S6, S7 enquires about the influence of qualified construction management on mandated overtime. Here, the time required for the detection of substandard workmanship is set to two or three days again.

S7.1—time required/detection of substandard quality 2 days: costs +5,58 %, time +6,87 %
S7.2—time required/detection of substandard quality 3 days: costs +7,84 %, time +13,00 %

5.1.1.8   Scenario S8: Crisis Management of a Bad Case Scenario

Considering all the above-mentioned scenarios, an exemplary bad case scenario is constructed and attempted to be balanced with the help of vocational adjustment effects (S2). Scenario S3 is used as a basis and combined with an average time of two days needed for the detection of substandard workmanship. Additionally, the S8 scenario assumes a continuous material loss of 10 % in the processing of concrete. This constellation leads to a massive exceeding of the contractually fixed maximum completion time. Integrating S2 can counteract this effect to a degree: costs +9,93 %, time +16,03 %.

5.1.2   Discussion of the Simulation Experiments

The contractually set completion time of 160 days is undercut by 18,12 % in the basic scenario S1. This means that in further scenarios a potential time delay would be tolerable up to this percentage—independently of the incurred costs. Based on the basic scenario S1, the scenario S2 with the inclusion of vocational adjustment effects leads to a reduction of the overall project costs of nearly 3 % and saved time of more than 8 % (Fig. 5.1).

It is questionable, however, in how far such effects of adjustment to the work can be initiated and operatively maintained for extended periods of time. Therefore, the result of S2 demonstrates the theoretical cost-saving potential of vocational adjustment effects, but further research is required in this field to certify how realistic these effects are. In contrast to S2, scenario S3 considers a reduced experience of the construction site personnel and its effect on the quality of realized workmanship. A reduction of 10 % results in additional overall project costs of almost 5 % and additional time requirements of 10 %. The increase of costs as

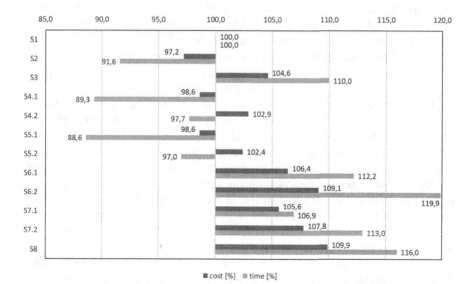

Fig. 5.1 Simulation results

well as needed time fulfilled the expectations. The fact that the percentage of the additionally needed time precisely matched the percentage of reduced experience in this scenario is a coincidence which can be proven by variations with other reductions of experience. A mandated overtime of 20 % leads to the reduction of overall costs of more than 1 % in scenario S4.1. Especially apparent is the saved time in this scenario, at almost 11 % a substantially higher value than in scenario 3 including the factor of work adjustment. Following the assumption that mandated overtime has a negative impact on the quality of work, scenario S4.2 displays an increase of the overall project costs of nearly 3 %. Despite the necessary cycles of revision and repair, project duration shrank by more than 2 %. This means that the influence on the quality of workmanship caused additional costs due to the cycles of revision and repair, but nevertheless the overtime still affected project duration in a positive manner. Especially in these scenarios, a valid determination of both increases and decreases in costs or required time is substantially difficult, as a reliable statement would need more detailed input data for the illustration of the relationship between overtime and work quality. The missing input data could be obtained from additional field studies. Despite that, the developed simulation model appears to depict the effects of the different scenarios correctly and can, therefore, demonstrate the relationships at least qualitatively. The scenario S5.1, originally seen as good case and supposed to show the advantages of combining overtime (S4.1) and adjustment to the work (S2), demonstrated a reduction of project costs indeed and a decrease of project time of almost 12 % in comparison to the basic scenario S1. However, when compared to the overtime scenario S4.1, the inclusion of work adjustment effects has only marginal impact on the final values. The reason for this is the availability of the required material. While the materials and resources necessary for the implementation of the first production cycle are available in sufficient quantity at the start of the scenarios S1, S3 and S4.1, the situation which develops in S5.1 does not allow the crew to deploy their full performance potential. Although the increased frequency of order placement allows the execution of work processes, but the storage capacity limited to a specific value does not provide enough room for an increased stockpiling. This, in turn, leads to a production process which is not running smoothly and in the best way. Even though a theoretically higher production performance is available, this is only applicable up to a certain limit. Thus the expected advantages cannot be realized in scenario S5.1, because the production performance is constricted by the storage situation. The scenario S5.1 demonstrates that no more than 12 % of time can be saved under the given circumstances. If the stockpiling is adapted accordingly, scenario S5.1 can realize a theoretical time saving of up to 41 %. Scenario S5.2 shows that the negative effects of mandated overtime in scenario 4.2 can be countered marginally only with the inclusion of vocational adjustment effects. Compared to the basic scenario S1, S5.2 results in a cost increase of more than 2 %. The overall project duration can be reduced by approximately 3 %. In addition to that, the results derived from the S5 scenarios highlight the fact that due to diverse interdependencies and feedback effects within the system, a superposition of results from different scenarios (e.g. S4+S2) is not directly possible.

A relationship between less experienced construction site personnel and the quality of the management can be extracted from the results of scenario S6. This scenario offers the expected result: the necessity of closer quality monitoring in case of less experienced workers appears generally reasonable. This problem is even more relevant when integrating services provided by subcontractors, as— in contrast to employed staff members—the construction management (especially at the beginning of the project) normally has no valid reference points for the qualification of the external staff. If substandard and flawed workmanship is continually only detected and remedied after the second day, the overall project costs increase by more than 6 % due to the reworking and repair. The project duration increases by more than 12 %. In case the shortcomings and dissatisfactory work results are continually detected and repaired after the third day, the overall project costs increase by 9 %, the completion time increases by nearly 20 % and nearly reaches the contractual maximum allowance of the project duration.

Similar, but less substantial consequences can be derived from the scenario S7, in which the time needed for the detection of flawed work results is inspected in combination with the reduced quality of work due to mandated overtime. In S7.1, the overall project costs rise by more than 5 %. Even though S7.1 leads to an increased project duration of nearly 7 %, this value is still below the corresponding value from S6.1. Therefore, the reworking and repair cycles are less time-consuming in this constellation due to the mandated overtime. If the time needed for detection of substandard work results is increased to three days (S7.2), the overall costs increase by almost 8 % in comparison to S1. The overall project duration is still within the contractual limits with an increase by 13 %.

The results of S6 and S7 clearly show the high relevance of quality management by the construction management. As quality and performance of the management depends on a diversity of factors, these results cannot be declared as final and comprehensive. Nevertheless, they hint at the impact and importance of late detection of flaws in workmanship. The bad case scenario depicted in S8, which is supposed to be balanced with the inclusion of the adjustment to work effect (S2), shows that the financial tolerance is nearly exceeded by 10 %. The originally calculated completion time is exceeded by almost 16 % and stays within the time limit set in the contract. If the material losses are increased or the detection period of substandard work result is longer than two days, the schedule of the project cannot be followed any more. In such a case, further parameters, e.g. personnel resources, would need to be adjusted accordingly. Even though the projects can be completed within the set time limit in the discussed scenarios and a (in German literature communicated) typical risk- and profit markup of between 1 %–4 % (Künstner et al. 2002; Hoffmann 2006) is assumed, a significant loss occurred in some of the projects. While not each project of a corporation is completed negatively and therefore the more successful projects balance the less successful ventures, the previous scenarios suggest that the usual risk- and profit markups appear to be insufficient and need to be reconsidered (one solution is suggested by Poloczek 2013).

References

Hoffmann M (2006) Beispiele für die Baubetriebspraxis, 1st edn. Vieweg+Teubner

Ibbs W, Vaughn C (2015) Change and the loss of productivity in construction: a field guide. Tech. rep., The Ibbs Consulting Group

Künstner G et al. (2002) Ablauforganisation von Baustellen - ein Leitfaden zur Planung und Steuerung von Bauabläufen am Beispiel einer Fertigungshalle mit Verwaltungsgebäude in Mischbauweise, 2nd edn. Wirtschaftliche und eektive Betriebsführung. Zeittechnik-Verl., Neu-Isenburg

Poloczek A (2013) Wertorientierte Bewertung von Projekten mit Unikat-Charakter. Ph.D. thesis, Institut für Baubetrieb und Baumanagement, Universität Duisburg-Essen

Chapter 6
Summary and Outlook

The aim was to introduce the development and practical testing of a module-oriented modeling approach, which can be used in the simulation of multi-causal and dynamic relationships on different levels of an industry. The conceptual design of such a development framework for domain-specific system-dynamic libraries (SDL approach) demonstrated that it is quite reasonable and possible to support the development of simulation models in that manner. Thereby, knowledge synergies are created which enable the interdisciplinary development of simulations in the sense of a synergistic knowledge absorption (named the SKA method).

This approach can be employed in an infinite number of applications. One example is the investigation of changes and their consequences within a project. Change may occur late for various reasons. One explanation is that some discrepancies, omissions, and needed work alterations are not discovered until the project is relatively far along (Ibbs 2005). In this case the introduced framework appears particularly helpful to investigate the project and its potential risks beforehand. Alternatively, this approach enables analyzing projects after completion to discover what and why happened in detail.

Based on this, strategies can be formulated that, e.g. ensure that fabrication and construction proceed while changes are being resolved (Ibbs and Backes 1994). The findings within the prototypical implementation can be used as indicative of general trends. Every industry, every project is different, so slavish reliance on theses data is not warranted.

In the future, multidisciplinary R&D teams of practitioners and scientists from different domains have numerous possibilities to develop SDL units from varying perspectives. Compared to other fields, significantly less implementations of simulations exist in the construction industry. Therefore, the approach introduced here can provide a valuable contribution to promote further developments especially in this domain.

© The Author(s) 2016
C.K. Karl, W. Ibbs, *Developing Modular-Oriented Simulation Models Using System Dynamics Libraries*, SpringerBriefs in Electrical and Computer Engineering, DOI 10.1007/978-3-319-33169-0_6

Beyond the formal procedure for the design of an SDL simulation, the first foundation of a domain-specific SDL (CDL) could be placed in the field of simulation. Apart from that, it was possible to demonstrate the practical implementation in the simulation program Vensim. Further, the synergistic modeling approach enables to explore and research human behavior—in addition to the classic use as a pure simulation method.

On the basis of the SDL approach respectively the domain-specific CDL presented here, an open database could be set up for the scientific community as well as for practitioners. The units and models presented here are available on www.constructiondynamics.org. This stock can successively be expanded by further developments from the community and exchanged on the above-mentioned website. The internet platform functions as a type of shared modeling platform (SMP) and allows the collaborative development of units and complete models. Apart from the exchange and the continuous expansion of the library, the encouraged discourse in the community leads to a validation and thereby quality management of the CDL units. Primary prerequisite is, however, an open, fair and forgiving culture of development, which is based on the common goal to create sufficiently detailed and valid units and models.

In this context, CDL units can be of potential interest in the future to examine the relationships between planning, execution and operation processes. The introduced system dynamics modeling approach could be beneficial as an addition in e.g. Building Information Modeling (BIM), Lean Construction (LC) Methods as well as in the operation and management of facilities.

Regarding the current trends in the field of sustainable construction, the previously developed units can be expanded with additional attributes, which allocate in the individual sub-processes the CO_2 emissions, for example (e.g. dependent on the power of the device or on the carbon footprint of individual building materials). This would facilitate to estimate the costs and ecological burden of projects in advance.

Furthermore, CDL units, especially those serving the illustration of risks in different areas of the industry, could offer the basis for the development of special risk or danger models which can assist in the investigation of specific situations or behaviors and later the configuration of precaution and safety measures.

This could be, for instance, the elucidation of the risk situation of a company, the identification and evaluation of project risks or endangered operational procedures on various functional levels.

In addition to this described application, the SDL approach appears to be helpful in other areas of modeling and simulation. Apart from the domain-specific SDL, www.systemdynamicslibrary.org is supposed to serve as a further platform on which domain-independent SD libraries and their units can be published and exchanged.

As explicitly different levels and functional areas of a company are considered, an application in the research of organizational psychology would be possible, too. The SDL approach and the contained units could support the generation of hierarchically structured models through different levels and areas. Hence, interaction between the different levels (single individuals, teams or other functional units) could be interlaced and examined under consideration of further methods like the multi-level analysis (MLA: For details on MLA refer to Ditton 1998, Raudenbush and Bryk 2002, Langer 2009, Hox 2010.). This means that SDL units and the resulting models can generally be examined with other simulation methods, too.

References

Ditton H (1998) Mehrebenenanalyse: Grundlagen und Anwendungen des hierarchisch linearen modells. Juventa, Weinheim

Hox JJ (2010) Multilevel analysis. Techniques and applications. Quantitative methodology series. Routledge, New York

Ibbs W (2005) Impact of change's timing on labor productivity. J Construct Eng Manag 131(11):1219–1223

Ibbs W, Backes JM (1994) Project change management. Tech. Rep. 43–1, Construction Industry Institute, Austin

Langer W (2009) Mehrebenenanalyse: Eine Einführung für Forschung und Praxis, 2nd edn. Verlag für Sozialwissenschaften/GWV Fachverlage GmbH, Wiesbaden

Raudenbush SW, Bryk AS (2002) Hierarchical linear models: applications and data analysis methods. Advanced Quantitative Techniques in the Social Sciences. Sage Publications, Thousand Oaks

Chapter 7
Appendix

7.1 Quantitative Data for the Simulation Experiments

- The simulation is based on an office building with the following boundary conditions:

  - Gross floor area $5.300\,m^2$
  - Three identical storeys
  - Reinforcement proportion: base plate and columns $0,13\,to/m^3$, ceilings $0,10\,to/m^3$
  - Base plate: $d = 0,30\,m$
  - Square columns: $h = 3,50\,m$, $0,25\,m^2$
  - Number of columns for each floor: 15
  - Ceilings: $d = 0,35\,m$

- The following devices are represented:

  - Tower crane (C.0.10.0071 cf. Baugeräteliste 2007) with trolley jib, load torque 71 tm, radius 35 m, machine performance 31 kW
  - Truck(D.6.00.1023 cf. Baugeräteliste 2007), load capacity 10.2 to, bowl content $5.5\,m^3$, machine performance 230 kW, approximate performance $290\,m^3$/day
  - Excavator(D.1.00.0060 cf. Baugeräteliste 2007), shovel capacity $0.6\,m^3$, machine performance 60 kW, approximate performance $288\,m^3$/day
  - Timber girder system formwork for ceilings (unit price for simulation including ceiling joist formwork: $10.00\,Euro/m^2$)
  - System formwork for columns (unit price for simulation: $30.00\,Euro/m^2$)

© The Author(s) 2016
C.K. Karl, W. Ibbs, *Developing Modular-Oriented Simulation Models Using System Dynamics Libraries*, SpringerBriefs in Electrical and Computer Engineering,
DOI 10.1007/978-3-319-33169-0_7

- soil disposal costs (Source: anonymized expert interview):

  – Soil LAGA Z.2 (limited installation with defined technical safety measures): 20.00 Euro/m^3
  – Transport distance between construction site and nearest disposal site for LAGA Z.2: 60 km

- Performance factors for various processes (Source: anonymized construction project (office building) in North Rhine-Westphalia/Germany):

  – Formwork

    * Ground plate, conventional formwork: 0.15 h/m^2
    * Columns system formwork: 1.15 h/m^2
    * Ceilings including ceiling joist, timber girder system formwork: 0.70 h/m^2

  – Reinforcement

    * Install reinforcing bars in columns: 21.00 h/to
    * Welded wire fabric in plate elements, cut and install: 17.00 h/to

  – Concreting

    * Reinforced foundation: 0.60 h/m^3
    * Columns, concreting with crane bucket: 1.30 h/m^3
    * Ceilings including ceiling joist: 0.70 h/m^3

- According to the federal collective agreement for the building industry (in German: Bundesrahmentarifvertrag für das Baugewerbe (BRTV)):

  – Weekly working hours 38 h (December to March) and 41 h (April to November)
  – Overtime bonus of 25 %
  – Employees as skilled workers (wage group 2)
  – Machinists as equipment operator or professional drivers (wage Group 3)

- Material prices (Concrete: Heidelberger Beton Donau-Naab GmbH & Co. KG, Steel: Ludwig Schierer GmbH, Timber: WIRBAU GmbH (data status 31st May 2013)):

  – Concrete C35/45 F3 (columns): 140.00 Euro/m^3
  – Concrete C30/37 F3 (ceilings), including concrete pump: 145.00 Euro/m^3
  – Welded wire fabric BST 500 M(A), ductility class A, BSTG Q 188 A to BSTG Q 424 A, including surcharge for support baskets: 2,000 Euro/to
  – Steel bars IV 500 S, 8 mm to 28 mm: 1,200 Euro/to
  – Wooden formwork spruce/pine, unplanned, waterproof: 6.00 Euro/m^2

- Market prices for operating supplies (data status 31st May 2013):

  – Diesel: 1.36 Euro/l, electricity 0.36 Euro per kWh

- Rental rates for containers from currently freely accessible databases (e.g. http://www.mietverbund.com):

  - Personal container (site manager, foreman, crew, sanitation, first aid, meeting room): 25.00 Euro/day
  - Material- and device-related containers (building materials magazine, devices magazine, operating supplies magazine): 10.00 Euro/day
  - Especially equipped container (laboratory, workshop): 50.00 Euro/day
  - Other costs as lump-sum per simulation step (electricity, water, telecommunications, etc.)

Furthermore, the following boundary conditions apply for the project simulation:

- Only the processes site preparation, earthworks, ground plate, columns and ceilings are considered.
- Three employees are always the minimum number in a working group.
- The performance factors can be influenced by a learning curve with a learning rate of 95 % and result in maximum productivity increase of +30 %.
- Motivation and experience of the workforce contribute equally weighted on the quality of workmanship. Sinking motivation and/ or experience of the workforce, which leads to poor workmanship quality, leads in turn to a revision cycle.
- According to the collective agreement overtime with max. 20 % of the working day can be realized.
- The overtime should influence the quality of workmanship.
- To represent the quality of the site management the time to detect poor workmanship can be varied between one and three days.
- A loss of material can be involved in all processes.
- Parallelization of processes is not included.
- Amendments by the contracting authority as well as storage losses are excluded.
- Orders are generally shipped out the next day. There are no further delays involved (e.g. by initiation and processing of the order).
- System formwork elements have no wear.
- All necessary storage space is available on the site surface. Therefore, costs by renting additional land for storage is excluded.
- All payments shall be made immediately without time delay.

7.2   Approximated Cost Functions

Resulting from the simulation, time-dependent cost curves are approximated as polynomials ($f : t \rightarrow at^n$ $n \in [5, 6]$, $t \rightarrow [0, t_{max}]$, t = time in days). The coefficient of determination is 0.969 to 0.985. Using the following equations, a good approximation to the simulation data can be achieved in the range of $t = 0$ and t_{max}. To avoid that the function is unstable, the decimal places indicated are important to consider (especially in terms of high potency).

7.2.1   Scenario S1: Basis

$$y_{S1}(t) = 7,697166 \cdot 10^{-4} t^5 - 0,2794371 t^4 + 37,23775 t^3 - 2212,084 t^2 + 61031,1t$$
$$\tag{7.1}$$
$$R^2 = 0,970, \; t_{max} = 131$$

7.2.2   Scenario S2: Vocational Adjustment

$$y_{S2}(t) = 1,17 \cdot 10^{-3} t^5 - 0,39 t^4 + 47,7 t^3 - 2600 t^2 + 65600t \tag{7.2}$$
$$R^2 = 0,974, \; t_{max} = 120$$

7.2.3   Scenario S3: Experience

$$y_{S3}(t) = 5,099658 \cdot 10^{-4} t^5 - 0,201043 t^4 + 29,09301 t^3 - 1878,209 t^2 + 56723,59t$$
$$\tag{7.3}$$
$$R^2 = 0,969, \; t_{max} = 144$$

7.2.4   Scenario S4: Overtime

$$y_{S4.1}(t) = 0,0013 t^5 - 0,4237 t^4 + 50,684 t^3 - 2699,1 t^2 + 66631t \tag{7.4}$$
$$R^2 = 0,975, \; t_{max} = 117$$
$$y_{S4.2}(t) = 8,694 \cdot 10^{-4} t^5 - 0,3074 t^4 + 39,87 t^3 - 2303 t^2 + 62080t \tag{7.5}$$
$$R^2 = 0,973, \; t_{max} = 128$$

7.2.5   Scenario S5: Overtime (S4) and Vocational Adjustment (S2)

$$y_{S5.1}(t) = 1,301 \cdot 10^{-3} t^5 - 0,4199 t^4 + 49,87 t^3 - 2651 t^2 + 66010t \tag{7.6}$$
$$R^2 = 0,973, \; t_{max} = 116$$

$$y_{S5.2}(t) = 8,993852 \cdot 10^{-4} t^5 - 0,3146805 t^4 + 40,45985 t^3 - 2322,295 t^2 + 62341,95t$$
$$\tag{7.7}$$
$$R^2 = 0,973, \; t_{max} = 127$$

7.2.6 Scenario S6: Experience (S3) and Variation of Time for Quality Management

$$y_{S6.1}(t) = -6,4244 \cdot 10^{-6}t^6 + 3,2742 \cdot 10^{-3}t^5 - 0,64667t^4 + 62,052t^3 - 2959,9t^2 + 68669t$$

(7.8)

$$R^2 = 0,982, \ t_{max} = 147$$

$$y_{S6.2}(t) = -5,2442 \cdot 10^{-6}t^6 + 2,7667 \cdot 10^{-3}t^5 - 0,5654t^4 + 56,129t^3 - 2770,3t^2 + 66656t$$

(7.9)

$$R^2 = 0,983, \ t_{max} = 157$$

7.2.7 Scenario S7: Overtime (S4) and Variation of Time for Quality Management

$$y_{S7.1}(t) = -9,6001 \cdot 10^{-6}t^6 + 4,5233 \cdot 10^{-3}t^5 - 0,82757t^4 + 73,814t^3 - 3288,8t^2 + 71756t$$

(7.10)

$$R^2 = 0,983, \ t_{max} = 140$$

$$y_{S7.2}(t) = -7,7242 \cdot 10^{-6}t^6 + 3,7929 \cdot 10^{-3}t^5 - 0,72253t^4 + 66,987t^3 - 3093,8t^2 + 69900t$$

(7.11)

$$R^2 = 0,985, \ t_{max} = 148$$

7.2.8 Scenario S8: Crisis Management of a Bad Case Scenario

$$y_{S8}(t) = -5,555 \cdot 10^{-6}t^6 + 2,891 \cdot 10^{-3}t^5 - 0,5836t^4 + 57,35t^3 - 2809t^2 + 67100t$$

(7.12)

$$R^2 = 0,981, \ t_{max} = 152$$

Reference

Baugeräteliste (2007) Baugeräteliste 2007. Bauverlag, Gütersloh

Chapter 8
Further Readings

Allwein G, Barwise J (eds) (1996) Logical reasoning with diagrams. Oxford University Press, New York

Barlas Y (1989) Multiple tests for validation of system dynamics type of simulation models. Eur J Oper Res 42(1):59–87

Börger E, Stärk R (2003) Abstract state machines: a method for high-level system design and analysis. Springer, Berlin

Burke WW (1994) Diagnostic models for organization development. In: Bray DW (ed) Diagnosis for organizational change: methods and models. The Guilford Press, New York, p 53–84

Carvalho GF (1995) Modelling the law of demand in business simulators. Simulat Gaming 26(1):60–79

Coyle RG (1977) Management system dynamics. Wiley, New York

Doets K (1996) Basic model theory. CSLI Publications, Stanford

Forrester JW (1989) The beginnings of system dynamics (Working Paper No. D-4165). System Dynamics Group, Sloan School of Management, MIT, Cambridge

Fowler M (2000) UML distilled. Addison-Wesley, Boston

Garnham A (2001) Mental models and the interpretation of anaphora. Taylor and Francis, Philadelphia

Gentner D, Stevens A (eds) (1983) Mental models. Lawrence Erlbaum, Hillsdale

Gigch van JP (1991) System design modelling and metamodelling. Plenum Press, New York

Hodges W (1997) A shorter model theory. Cambridge University Press, Cambridge

Homer JB (1983) Partial-model testing as a validation tool for system dynamics. In: International System Dynamics Conference, p 920–932

Lyneis JM (1980). Corporate planning and policy design. Productivity Press, Cambridge

Manzano M (1999) Model theory. Oxford University Press, Oxford

Mass NJ (1975) Economic cycles: an analysis of underlying causes. Productivity Press, Cambridge

Meadows DH, Meadows DL, Randers J, Behrens III WW (1972) The limits to growth: a report for the Club of Rome's project on the predicament of mankind, Universe Books

Meadows DH, Meadows DL, Randers J (1992) Beyond the limits. Global collapse or a sustainable future. Earthscan Publications, London

Meadows DL (1970) Dynamics of commodity production cycles. Productivity Press, Cambridge

Meadows DL, Meadows DH (eds) (1974) Toward global equilibrium: collected papers. Productivity Press, Cambridge

Meadows DL, Behrens III WW, Meadows DH, Naill RF, Randers J, Zahn EKO (1974) Dynamics of growth in a finite world. Productivity Press, Cambridge

© The Author(s) 2016

C.K. Karl, W. Ibbs, *Developing Modular-Oriented Simulation Models Using System Dynamics Libraries*, SpringerBriefs in Electrical and Computer Engineering, DOI 10.1007/978-3-319-33169-0_8

Merten PP (1991) Loop-based strategic decision support systems. Strat Manag J (12):371–382

Meijers A (ed) (2009) Philosophy of technology and engineering sciences. Elsevier, Amsterdam; see chapters W. Hodges, Functional modelling and mathematical models; R. Müller, The notion of a model, theories of models and history; and N. Nersessian, Model based reasoning in interdisciplinary engineering

Morecroft JDW (2007) Strategic modelling and busines dynamics. Wiley, Chichester

Morecroft JDW (1985) Rationality in the analysis of behavioral simulation models. Manag Sci 31(7):900–916

Pullum GK, Scholz BC (2001) On the distinction between model-theoretic and generative-enumerative syntactic frameworks, in Logical Aspects of Computational Linguistics (Lecture Notes in Computer Science: Volume 2099), P. De Groote et al. (eds), Springer, Berlin, p 17–43

Richardson GP (1986) Problems with causal-loop diagrams. Syst Dynam Rev 2(2):158–170

Roberts EB (ed) (1978) Managerial applications of system dynamics. Productivity Press, Cambridge

Rothmaler P (2000) Introduction to model theory. Gordon and Breach, Amsterdam

Stenning K (2002) Seeing reason. Oxford University Press, Oxford

Sterman J (2002) All models are wrong: reflections on becoming a systems scientist. Syst Dynam Rev 18(4):501–531

Sterman J (1989) Misperceptions of feedback in dynamic decision making. Organ Behav Hum Decis Process 43(3):301–335

Suppes P (1969) Studies in the methodology and foundations of science. Reidel, Dordrecht

Warren K (2002) Competitive strategy dynamics. Wiley, Chichester.

Index

© The Author(s) 2016   103
C.K. Karl, W. Ibbs, *Developing Modular-Oriented Simulation Models Using System
Dynamics Libraries*, SpringerBriefs in Electrical and Computer Engineering,
DOI 10.1007/978-3-319-33169-0